ENDOMYOMETRIAL PATHOLOGY – A BRIEF OVERVIEW

Dr. Gunvanti Rathod

Dr. Pragnesh Parmar

ENDOMYOMETRIAL PATHOLOGY – A BRIEF OVERVIEW

Dr. Gunvanti Rathod, MD (Pathology)

Assistant Professor
Department of Pathology
SBKS Medical Institute and Research Centre
Vadodara, Gujarat, India

Dr. Pragnesh Parmar, MD (Forensic Medicine)

Assistant Professor
Department of Forensic Medicine
SBKS Medical Institute and Research Centre
Vadodara, Gujarat, India

DEDICATION

This book is dedicated to my loving sister **Dr. Sangita Rathod**, her husband **Dr. Ashish Parikh** and her daughter **Jaladhi**.

- **Dr. Gunvanti Rathod**

ACKNOWLEDGEMENTS

We acknowledge the immense help received from the scholars whose articles are cited and included in references of this book. The authors are also grateful to authors / editors / publishers of all those articles, journals and books from where the literature for this book has been reviewed and discussed.

We express our gratitude to our parents and in-laws for their constant encouragement, support and blessings.

It will be an injustice if we do not thank all our students for their innovative ideas and feedback.

CONTENTS

TOPIC	PAGE NUMBER
PATHOLOGY OF ENDOMETRIUM	
INTRODUCTION	7
ANATOMY OF ENDOMETRIUM	9
HISTOLOGY OF ENDOMETRIUM	11
MENSTRUATION AND OTHER CYCLICAL PHENOMENA	14
ABNORMAL UTERINE BLEEDING	29
ENDOMETRITIS	45
ENDOMETRIAL POLYPS	51
ENDOMETRIAL INTRAEPITHELIAL NEOPLASIA (EIN)	68
CARCINOMA OF ENDOMETRIUM	71
GESTATIONAL TROPHOBLASTIC DISEASES	93
MYOMETRIAL PATHOLOGY	
INTRODUCTION	98
EMBRYOLOGY	99
ANATOMY	100
HISTOLOGY	101
CLASSIFICATION	102
LEIOMYOMA	104
SMOOTH MUSCLE TUMOR OF UNSUAL MALIGNANT POTENTIAL	124

(STUMP)	
LEIOMYOSARCOMA	125
MISCELLANEOUS TUMORS OF MYOMETRIUM	128
REFERENCES	132

PATHOLOGY OF ENDOMETRIUM

INTRODUCTION

- The endometrium which lines the uterine cavity is one of the most dynamic tissues in the human body; it is one of the interesting tissues for histopathologic study. It is characterized by cyclic processes of cell proliferation, differentiation and death in response to sex steroids elaborated in the ovary. An understanding of the varieties in the normal morphological appearance of the endometrium provides an essential background for the evaluation of endometrial pathology. [1]

- Abnormal uterine bleeding is one of the most common problems in all age groups. The abnormal bleeding can be caused by a wide variety of disorders. It may represent a normal physiological state, and observation alone may be warranted. Alternatively, the bleeding can be a sign of a serious underlying condition necessitating aggressive treatment that could include a major procedure. [2]

- The causes for the bleeding in elderly women are hormonal, pregnancy complications, bleeding diathesis and more importantly local pathology including malignancy, benign tumours, and infection. While Dysfunctional Uterine Bleeding (DUB) is responsible for most cases of abnormal uterine bleeding in the adolescent age group, the incidence of structural pathology increases in other age groups. [3]

- The bleeding in the perimenopausal period may be secondary to estrogen withdrawal (physiological state). In some cases it may be due to malignancy of the reproductive organs, particularly in postmenopausal women.

- The need to embark upon a diagnostic curettage in perimenopausal women cannot be overemphasized. [3]

- It is now generally accepted that an adequate clinical examination of abdomen and pelvis, and uterine curettage, hysteroscopy or at least an endometrial biopsy are essential to exclude organic disease of the uterus in these women. [4]

ANATOMY OF ENDOMETRIUM

- Endometrium, the lining mucosa of the uterus is a labile tissue, hormonally, responsive to sex steroids, elaborated in the ovary. In the first half of the menstrual cycle, all its elements undergo proliferation under the influence of estrogens, and in the later half, it responds to progesterones by the production of secretions and stromal alterations necessary for implantation of a fertilized ovum. [1]

- The endometrium at birth is characteristically thin and consists of a continuous surface of cuboidal epithelium which occasionally dips down to line a few small sparse tubular glands. The endometrial stroma is a thin definitive layer separate from the myometrium. The pre-menarchal production of estrogens promotes the growth of endometrial glands and stroma. [1]

- The adult uterus measures 50 - 80 grams. It measures 7 - 9 cm in length, 4.5 – 6 cm in width and anteroposterior thickness of 4 cm. The endometrium varies from 1-8 mm thickness and the myometrium averages 1.5 – 2.5 cm thickness. [5]

- During menopause the endometrium becomes thin and inactive, because of the failure of the ovary to respond to gonadotrophic hormones. Inactive endometrium becomes thin and may measure 1mm in thickness. [1]

Blood supply

- Main blood supply to the endometrium is by branches of uterine artery which arises from the anterior trunk of internal iliac or hypogastric artery. It reaches the uterus at the level of internal os, where it turns upwards at right angles and follows a spiral course along the lateral border of the uterus. During the vertical part of its course, branches which run transversely pass into the myometrium. These are called arcuate arteries and from them arise a series of radial arteries almost at right angles. From these are derived the terminal spiral and straight arterioles of the endometrium, which supply the functional zone. Two types of arteries supply the endometrium; one of these is restricted to the basal third and consists of small straight and short arteries. The superficial two thirds of the endometrium are supplied by coiled arteries. The veins follow the arteries and drain into the pelvic plexus of veins. [1]

Lymphatics

- The lymphatic's from the basal layer of the endometrium run through the myometrium in close relation to the blood vessels to reach the subserosal plexus. In the course of their passage through the parametrum and ovarian ligament, they communicate with the ovarian lymphatics to terminate in the pre aortic lymph nodes. [1]

- The lymphatic's from the lower uterine segment anastomose with adjacent lymph channels from the cervix and drain to the obturator, iliac, hypogastric and sacral nodes. [5]

HISTOLOGY OF THE ENDOMETRIUM

The endometrium of the corpus is composed of 2 layers:
1) The basalis, the layer from which the endometrium regenerates after menstrual shedding.
2) The overlying functionalis [1]

- In the second half of the menstrual cycle, the functionalis may be differentiated into the superficial compactum and the underlying spongiosa, which extends to the basalis. [1]

- During the menstrual cycle, the endometrium varies from 1 mm (post menstrual) to 8mm at the end of the third week. Every layer consists of 2 major components, the epithelial component, either as glands or as superficial epithelium and the mesenchymal component of stromal cells with pleuripotential properties. [1]

A) The glandular epithelium:

- It is a single layer of columnar epithelial cells. Their height varies depending on the functional (hormonal) state from 6 μm postmenstrual to 20 μm at the end of the proliferative phase.

- During the proliferative phase, the nuclei of the glandular cells are elongated and have a dense chromatin. During the secretory phase, the nuclei become round, vesicular and gradually lose their DNA.

- The apical surface of the epithelial cells is the proliferative phase; possess elongated delicate microvilli which contain alkaline phosphatases. During the secretory phase, as these microvilli draw back and disappear, the activity of alkaline phosphatase diminishes.

- It is also possible to find ciliated cells among the glandular epithelial cells. The ciliated cells initially lie against the basement membrane and because of their abundant translucent cytoplasm they can readily be recognized as 'clear cells'. They have a rounded nucleus which is located above those of the neighbouring epithelial cells.

- Each cell has eleven cilia's. The number of ciliated cells fluctuates considerably from patient to patient depending on the functional state of the endometrium. The ciliated cells are more abundant close to the tubal cornua and the endocervical mucosa.

B) <u>**The superficial epithelium:**</u>

- It contains greater number of ciliated cells than the glandular epithelium during the proliferative phase.

C) <u>**The stromal cells:**</u>

- The endometrial stroma consists of pleuripotential mesenchymal cells, which at the beginning of the menstrual cycle are uniformly spindle shaped and poorly differentiated which are joined to one another by cytoplasmic processes. The cells lie firmly anchored within a delicate network of reticulum fibres.

- Characteristically it is spindle shaped with scant cytoplasm surrounding a fusiform nucleus. In addition to fibroblasts, leucocytes, mast cells and plasma cells are frequently found in the stroma.

D) <u>**The reticulum fibres:**</u>

- The reticulum fibres in contrast to collagen fibres may be reformed within a few days giving rise to a dense reticular network. While the stroma of the basalis and isthmic mucosa remains uniformly dense, the content of fibres in the stratum functionalis fluctuates considerably during the menstrual cycle.

- Only occasional delicate reticulum fibres can be made out during the first 8 days of the proliferative phase. As ovulation approaches, these fibres became lesser and thicker. During the secretory phase, they are temporarily pulled apart by the transitory edema that develops. By 4th week of the cycle, they enmesh each predecidual cell and form a dense network around the glands and spiral arterioles. With decrease in progesterone level the reticulum fibres disintegrate.

E) The vessels:

- The vessels of the stratum functionalis of the endometrium are very much sensitive to hormones where as the vessels of the basalis are influenced little by hormonal changes of the cycle. There is an irregular network of venous channels with the veins frequently intersecting forming venous lakes. [5]

MENSTRUATION AND OTHER CYCLICAL PHENOMENA

- Menstruation is defined as a 'periodic and cyclical shedding of progestational endometrium accompanied by loss of blood' during the reproductive age between the menarche (onset of menstruation) and menopause (cessation of menstruation). [7]

- The normal menstrual cycle takes place at approximately 28 days intervals with a range of 21 – 35 days, the flow lasts 4 ± 2 days, and the average blood loss is 40 ± 20 ml.

- The menstrual cycle activity is under complex hormonal control and endometrial morphology closely reflects the endocrine status and the interplay between ovarian hormones. Endocrine cyclic activity of the ovary and the hypothalamic-pituitary axis determines the different phases of a normal menstrual cycle. [8]

The normal human menstrual cycle can be divided into two segments
- The ovarian cycle
- The uterine cycle
- The ovarian cycle is further divided into:
- Follicular phase
- Ovulation
- Luteal phase

    - The uterine cycle is divided into corresponding proliferative and secretary phases. [9]

Four major hormones are involved in the control of menstrual cycle and can be measured in peripheral blood.
- Follicular stimulating hormone (FSH)
- Luteinizing hormone (LH)
- Estrogen, and
- Progesterone

Their secretion pattern is closely interrelated and reflects the cyclic patterns of hypothalamic activity.

Ovarian cycle

1. The follicular phase:

- During the follicular phase an orderly sequence of events takes place that ensures that proper number of follicles is ready for ovulation. In the human ovary the end result of this follicular development is (usually) one surviving mature follicle. This process, which occurs over the space of 10-14 days, features a series of sequential actions of hormones and autocrine/ paracrine peptides on the follicle, leading the follicle destined to ovulate through a period of initial growth from a primordial follicle through the stages of the preantral, antral and preovulatory follicles. [10] Variability in length of follicular phase is responsible for most variations in total cycle length. (Normal 10-14 days). [9]

The primordial Follicle

- The primordial germ cells originate in the endoderm of the yolk sac, allantois, and hindgut of the embryo, and by 5 to 6 weeks of gestation they migrate to genital ridge. [10]

- Multiplication of germ cells begin at 6 to 8 weeks of pregnancy and by 16 to 20 weeks maximum number of oocyte is reached (6 to 7 million in Both ovaries).

- The primordial follicle is non growing which consists of an oocyte surrounded by a single layer of spindle shaped granulosa cells. The first visible Signs of follicular development are an increase in the size of the oocyte and the granulosa cells becoming cuboidal in shape. [10]

A) Events in the preantral follicle:
- Initially development of follicles occurs independently of hormonal influence.
- This is followed by FSH stimulation which propels follicles to the preantral stage.

- FSH induces aromatization of androgen in the granulosa resulting in the production of estrogen. [10]
- And together, FSH and estrogen causes increase in the FSH receptor content of the follicle.

B) Events in the antral follicle:
- Estrogen production in follicular phase is explained by the two-cell, two gonadotropin mechanism.
- During days 5 to 7 dominant follicle selection is established and peripheral levels of estradiol begin to rise significantly by cycle day 7 consequently.
- And estradiol levels, which are derived from the dominant follicle, increase steadily and, through negative feed back effects, exert a progressively greater suppressive influence on FSH release.
- Due to a decline in FSH levels, the midfollicular rise in estradiol exerts a positive feedback influence on LH secretion.
- The positive action of estrogen also modifies the gonadotropin molecules, by increasing the quality as well as the quantity of FSH and LH at midcycle.
- Then LH levels rise steadily during the late follicular phase and stimulate androgen production in the theca.
- The dominant follicle utilizes the androgen as substrate and further accelerates estrogen production.
- LH receptors start appearing on granulosa cells by the action of FSH.
- Granulosa cells secrete inhibin B in response to FSH. This can suppress pituitary FSH secretions directly.
- FSH secretion and action is further augmented by activin (originating in pituitary and granulosa). [10]

C) Events in the Preovulatory follicle:
- Estrogen production is sufficient to achieve and maintain peripheral threshold concentration of estradiol which is required in order to induce the LH surge.
- Luteinization and progesterone production in the granulosa layer is initiated by LH through its receptors.
- This preovulatory rise in progesterone facilitates the positive feedback action of estrogen and induces the mid cycle FSH peak.

2. Ovulation:

- The preovulatory follicle, through the elaboration of estradiol, provides its own ovulatory stimulus. Considerable variation in ovulation timing exists from cycle to cycle, even in the same woman. [10]

- A reasonable and accurate estimate places ovulation approximately 10 - 12 hours after the LH peak and 24 - 36 hours after peak estradiol levels are attained. The onset of the LH surge appears to be the most reliable indicator of impending ovulation, occurring 34 - 36 hours prior to follicle rupture. A threshold of LH concentration must be maintained for 14 - 27 hours in order for full maturation of the oocyte to occur. Usually LH surge lasts 48 - 50 hours. [10]

Ovulatory events:

- The LH surge is responsible for continuous reduction division in the oocyte, luteinization of the granulosa, and for the synthesis of progesterone and prostaglandins within the follicle. [10]

- This progesterone and prostaglandins together enhance the activity of proteolytic enzymes which is responsible for digestion and rupture of the follicular wall. [10]

- The progesterone influence mid cycle rise in FSH is responsible to free the oocyte from follicular attachments and, to convert plasminogen to the proteolytic enzyme, plasmin, and to ensure that sufficient LH receptors are present to allow an adequate normal luteal phase.

3. Luteal phase:

- Before rupture of the follicle and release of the ovum, the granulosa cells begin to increase in size and assume a characteristic vacuolated appearance associated with the accumulation of a yellow pigment, LUTEIN, which lends its name to the process of luteinization and the anatomical subunit the corpus luteum. Angiogenesis is an important feature of the luteinization process.

- Luteal phase - the time from ovulation to the onset of menses, with an average length of 14 days.

Events in the luteal phase:

- Adequate FSH stimulation and continued tonic LH support is required for normal luteal function. [10]

- New follicular growth is suppressed by progesterone which is acting centrally and within the ovary.

- Corpus luteum regression involves the luteolytic action of its own estrogen production, mediated by an alteration in local prostaglandin and endothelin- I concentrations. [10]

- In early pregnancy, the corpus luteum is rescued by human chorionic gonadotropin (HCG) which maintains luteal function until placental functions develop. [10]

The luteal- follicular transition:

- The interval extending from the late luteal decline of estradiol and progesterone production to the selection of the dominant follicle is a critical and decisive time, marked by the appearance of menses, very important are the hormone changes that initiate the next cycle. The critical factors include GnRH (Gonadotropin releasing hormone), FSH, LH, estradiol, progesterone, and inhibin. [10]

Events in the luteal follicular transition:
- The demise of the corpus luteum results in a nadir in the circulating levels of estradiol, progesterone, and inhibin.

- FSH secretion suppressing influence in the pituitary is removed by the decreasing levels of inhibin A. [10]

- Also the decrease in estradiol and progesterone will allow a progressive and rapid increase in the frequency of GnRH pulsatile secretion and removal of the pituitary from negative feedback suppression. [10]

Uterine cycle: [10, 11, 12, 13]

The mucosal lining of the uterus is composed of the glands and the stroma. The endometrium of the corpus is composed of two layers,

1. **The basalis:** is the deepest region of the endometrium and does not undergo significant monthly proliferation Instead, it is the layer from which the endometrium regenerates after shedding.
2. **Overlying Functionalis:** It is the superficial two thirds of the endometrium that proliferates and is ultimately shed with each cycle if pregnancy does not occur. In the second half of the menstrual cycle, the functionalism may be differentiated into superficial **compacta** and the underlying **spongiosa,** which extends to the basalis. [6]

The endometrium varies in thickness over the cycles. [11]
- At menstruation - 0.5mm thick
- Immediate post menstrual phase -1-2mm
- Proliferative phase - 2- 4mm
- Mid secretory phase - 7-8mm

There is some reduction to 5-6mm in the immediate premenstrual phase. Customarily the normal menstrual cycle (in uterus) is divided into two main phases.
1. The proliferative phase
2. The secretory phase

To which can be added the menstrual phase.

- Estrogen predominates in the proliferative phase, the progesterone effects prevails in the secretory phase.

- There are nine histologic features of the glands and stroma that determine the phase of the cycle and the histologic date.

- Five of these features affect glands: (1) tortuosity, (2) gland mitoses, (3) orientation of nuclei (pseudostratified versus basal), (4) basal subnuclear cytoplasmic vacuoles, and (5) luminal secretions with secretory exhaustion. Four features relate to the stromal: (6) edema, (7) mitoses, (8) predecidual change and (9) infiltration of granular leucocytes. Practically, the most important glandular features are orientation of nuclei, subnuclear cytoplasmic vacuoles, and luminal secretions with secretory exhaustion (3, 4, and 5), and the most important stromal features are edema, predecidual change, and granular leucocytic infiltration (6, 8, and 9). These salient features are usually readily apparent when present, allowing the pathologist to assign a histologic date.

Morphologic features used in endometrial dating

- Glandular changes
  1. Tortuosity
  2. Mitoses
  3. Orientation of nuclei (pseudostratified or basal)
  4. Subnuclear cytoplasmic vacuoles
  5. Secretory exhaustion (luminal secretions)
- Stromal changes
  1. Edema
  2. Mitoses
  3. Predecidua
  4. Granular leukocytic infiltrate

1. The normal proliferative Phase: [12]

This phase generally l-asts two weeks but under physiological conditions may fluctuate between ten and twenty days. This phase is subdivided into early, middle and late stage.

A) The early proliferative stage: (Fourth to seventh day of a twenty eight day cycle)

- This phase is characterized by a low endometrium and represents essentially a freshly epithelialized basalis; glands are sparse, narrow and straight, embedded in a loose stroma of spindly cells. The epithelial cells of the glands are low columnar with little cytoplasmic ribonucleic acid (RNA). Their nuclei appear small, oval and the chromatin dense. Nucleoli are inapparent. The spindle stroma cells are well anchored in the reticulum network. The formation of their nuclei is dense and cytoplasm is scanty. As the effect of estrogen steadily intensifies, the endometrium gradually changes and comes to the mid proliferative stage.

B) Mid proliferative stage: (eight to tenth day of a twenty eight day cycle)

- The prime characterizing change in this stage is great increase in the height of the endometrium resulting from generalized stroma edema induced by estrogen. The glands start to become tortuous and long. Their epithelial glands become tortuous and long .Their epithelial cell becomes compressed and tall columnar with large, oval nuclei with dense chromatin.

- Nucleoli are apparent. Many cells show mitosis. The stroma is made up of spindle shaped cells with scanty cytoplasm and large fusiform nuclei, separated by the interstitial edema and lie attached to the reticulum network. Stromal cells in mitosis are abundant.

C) Late proliferative stage: (Eleventh to fourteenth day of a twenty eight day cycle)

- This phase is marked by the regression of edema. The endometrium temporarily shrinks and as a result the glands become more tortuous and their epithelial cells lining show a pseudostratified appearance. The apical edges of the cells of are sharp and smooth and it appears as if the lumina of the glands have been punched out. The lumina of the glands are either empty or contain scanty substance composed of proteins and mucopolysaccharides shed by the cells. The nuclei are large and fusiform with many small nucleoli, which become prominent with acridin orange flourochromation. Simontanously tiny granules of glycogen appear at the basal part of glandular cells, granules stain red with periodic acid Schiff (PAS) stain. The stroma is compact with large and proliferated stroma cells with prominent nucleoli.

Proliferative phase changes

- ❖ Early (4–7 days)
  - Thin regenerating epithelium
  - Short narrow glands with epithelial mitoses
  - Stroma compact with mitoses (cells stellate or spindle shaped)
- ❖ Mid (8–10 days)
  - Long, curving glands
  - Columnar surface epithelium
  - Stroma variably edematous, mitoses frequent
- ❖ Late (11–14 days)
  - Tortuous glands
  - Pseudostratified nuclei
  - Moderately dense, actively growing stroma

2. The normal secretory phase: [11, 12, 13]

- The normal secretory phase of most cycles lasts fourteen days (+1). The precise duration of the secretory phase is because, after ovulation the corpus luteum develops and involutes at a definite rhythm in a precise sequence, causing changes in the endometrium to take place at the same rate. Grossly secretory endometrium is 3 to 5 mm thick and creamy yellow in appearance.

- Morphological changes in the endometrium are not apparent until 36 - 48 hours after ovulation has occurred, so that the earliest evidence of secretory activity is not seen until the second postovulatory day. At this stage, there is an overlap of proliferative and secretory activity; the endometrial glands show both mitotic and secretory activity. This phase is divided into early, mid and late secretory phases.

- The first sign of secretion in the endometrium is the presence of sub nuclear vacuolation. The appearance of the subnuclear vacuolations holds the start of the early secretory phase.

A) Early secretory phase: Lasts from second postovulatory day to fifth postovulatory day. This phase is marked by prominent basal vacuoles pushing the nucleus towards the lumen of glandular cells. Glands become more tortuous but there is no longer the pseudostratified appearance of the epithelium.

B) Mid Secretory phase: Lasts from fifth day after ovulation until the eleventh day and is the longest subdivision within the menstrual cycle. The outlines of glands are markedly irregular, in contrast to round or oval pattern seen in the Proliferative phase.

- The secretory vacuoles are present at the luminal side of the glandular cells .The apical surface of the cell is rough and indistinct because of the Process of apocrine secretion and vacuoles are seen in the cytoplasm of the luminar part of the cell. Nuclei of epithelial cells are arranged in linear cushion and they are round, vesicular and pale staining up until about the seventh post ovulatory day, the stroma shows very little response to ovulation but from that time to the end of the cycle the stomal changes are more striking those in the glands. On the seventh day after ovulation stroma begins to show edema (peak at nine and ten days after ovulation). Stratification of the endometrium is more apparent. Groups of spiral arterioles become prominent ninth day after ovulation; they grow larger, thicker and become spirally twisted compared to proliferative phase. Stromal cells surrounding these spiral arterioles are large, round with increased RNA content. Ten days after ovulation the edema starts to regress and at the same time stroma cells start to become decidualized. This change is first appreciated as a solidification of the stromal cells around the spiral arterioles in contrast to the loose edematous patterns of the surrounding areas. Eleven days after ovulation, the stromal edema is reabsorbed and decidual change of the stromal cells is more marked and wide spread. The absorption of the stromal edema is the result of falling levels of estradiol and progesterone, which in turn are due to lysis of the corpus luteum.

C) Late secretory phase: is characterized by compact stroma without edema; the stroma of superficial zone show decidual change and many endometrial granulocytes. Stroma in the spongy zone is undifferentiated. In the central, the spongy part of the endometrium the glands have a characteristic 'saw toothed' appearance. Epithelial cells are moderately tall columnar and with full secretion. The glands of stratum compactum are few in number and lined by flattened atrophic looking cells. As the end of cycle approaches, the edema of the stroma regresses completely and decidualisation spreads throughout most of the endometrium and focal necrosis and haemorrhage occur in the stroma.

Endometrial dating of secretory phase

- ❖ Interval phase, 14–15 days. No datable changes for 36–48 hours after ovulation.
- ❖ Early secretory phase, 16–20 days. Glandular changes predominate

16 day	Subnuclear vacuoles (Note: Scattered small irregular vacuoles can becaused by estrogen alone.)
17 day	Regular vacuolation—nuclei lined up with subnuclear vacuoles
18 day	Vacuoles decreased in size Early secretions in lumen Nucleus approaches base of cell.
19 day	Few vacuoles remain. Intraluminal secretion No pseudostratification, no mitoses.
20 day	Peak of intraluminal secretions

- ❖ Mid- to late secretory phase, 21–27 days. Stromal changes predominate, variable secretory exhaustion

21 day	Marked stromal edema
22 day	Peak of stromal edema—cells have "naked nuclei"
23 day	Periarteriolar predecidual change Spiral arteries prominent
24 day	More prominent predecidual change Stromal mitoses recur
25 day	Predecidual differentiation begins under surface epithelium. Increased numbers of granular leukocytes
26 day	Predecidua starts to become confluent
27 day	Granular lymphocytes more numerous Confluent sheets of predecidua Focal necrosis
24 - 27 days	Secretory exhaustion of glands—tortuous with intraluminal tufts (saw-toothed), ragged luminal borders, variable cytoplasmic vacuolization, and

lumenal secretions

The menstrual phase: [13]

- If pregnancy has not occurred, the late secretory phase leads inevitably to menstrual phase, starts 14 days after ovulation. This phase is recognized histologically by crumbling of the stroma and glandular collapse and haemorrhage in the superficial stroma. On the second day of menstruation- scattered stromal cells and remnants of glandular epithelium lying amid fresh blood and aggregates of neutrophils are present.

Regeneration:

- Immediately after menstrual shedding ceases and before proliferation begin, regenerative phase sets in, lasting one to two days, during which the denuded endometrium becomes epithelized.

Theories on etiology of menstruation

- Systemic endocrine factors plays an undeniable role in the initiation of menstruation local factors is of importance in determining the origin of normal or DUB. Various theories explain the origin of uterine bleeding. [8]

1. Estrogen deprivation:
- Withdrawal of estrogen alone in the form of bilateral oophorectomy, destruction of mature follicles or discontinuation of estrogen therapy in a castrated subject follows endometrial disintegration and bleeding. Halving of the estrogen dose also follows bleeding conversely; discontinuation of estrogen treatment does not always produce uterine bleeding in post menopausal women. So it is evident that factors other than estrogens are involved in the pathogenesis of menstruation.

2. Progesterone deprivation:
- In ovulatory cycles, regression of corpus luteum causes menstruation. Endometrial shedding can be produced experimentally by removal of an active corpus luteum or by discontinuing progesterone alone after the simultaneous administration of estrogen and progesterone. Progesterone

withdrawal bleeding occurs only in an endometrium primed by estrogen, but however endometrial bleeding may also occur in the absence of progesterone (i.e., an ovulatory cycle). So other factors are operative in the initiation of uterine bleeding too.

3. **Inadequate lymphatic Drainage:**
- **Reynolds** (1947) suggests that the endometrial lymphatic system is inadequate to provide for removal of the products of catabolism brought about by local stasis, edema and leukocytic infiltration that result from withdrawal of endometrium metabolic support. The products of catabolism accumulate until the endometrium is shed down to the level of the basal arterioles.

4. **Menstrual toxin and the endometrial bleeding factor:**
- "**Smiths**" demonstrated the presence of an active euglobulin with pyrogenic, fibrinolytic and markedly toxic properties in menstrual blood, and in fresh endometrium just prior to menstruation. According to "**Markee**" (1950) this factor is a potent vasoconstrictor and becomes effective on withdrawal of sex steroids. It affects the spiral arterioles supplying the superficial layers of the endometrium. Vasoconstriction leads to ischemia, necrosis of the vessels and bleeding. The straight arteries supplying the basal endometrium do not respond to the action of this vasoconstrictive substance and allow the regeneration of endometrium.

- Prostaglandin $F_2\alpha$ - a group of substance related to fatty acids with vasoconstrictive and muscle stimulating and other properties, when infused in human volunteers induced menstruation.

5. **Depolymerization theory:**
- During the proliferative and early secretory phases, under the sequential and combined influence of estrogen and progesterone, enzymes like exopeptidases, acid phosphatase and other proteases, are manufactured and stored in the Golgi Lysosomal complex of endometrial cells. If implantation does not occur, progesterone production decreases, and

lysosome allows the release of these enzymes which are then responsible for endometrial breakdown.

Mechanism of normal menstruation

- In normal menstruation one half to three quarters of the menstrual discharge is blood rest being fragments of endometrial tissue and mucus. Menstrual blood does not clot normally and consists of aggregation of endometrial tissue, red cells and degenerated platelets and fibrin. Following are the changes seen in normal menstruation. [14]

1. Changes in the endometrium:
- The unique feature is the existence of spiral arteries or arterioles in the endometrium. In the proliferative phase the spiral arterioles grow upward from basal to more superficial layers of the endometrium. In the luteal phase there is marked increase in length and coiling of the arteriole with dilatation. In premenstrual phase the endometrium is shrinked and spiral arterioles become more coiled. At the same time gaps start appearing between endothelial cells of spiral arterioles and associated thin walled veins and leukocyte migrate through the gaps into the stroma. Immediately before menstruation the spiral arteriole starts constricting intensely for a period of 24 hrs and then dilates with a massive extravasation of erythrocytes into the stroma of the endometrium. According to Markee key event in menstruation is the vasoconstriction of the spiral arterioles due to the liberation of an unknown substance the endometrium which produces vasoconstriction, resulting in damage to the wall of the spiral arterioles and necrosis of the superficial layers the endometrium.

2. Haemostasis and endometrial regeneration:
- Primary haemostasis in the spiral arterioles is achieved by the formation of plugs of aggregated platelets and fibrin. After 24 hours the main mechanism ensuring haemostasis is constriction of the spiral arterioles, with swelling of the endothelial cells which completely occlude the spiral arterioles. On the second day of bleeding re-epithelialization commences from the basal glands and usually gets completed by third or fourth day. Rate of re-epithelialization depends upon the amount of estrogen

stimulation which is dependent on the rate of growth of the follicles developing in the ovary.

3. **Fibrinolysis and liquefaction of menstrual blood:**
   - Liquefaction of menstrual blood is an important part of the mechanism of menstruation, not only in facilitating the passage of the menstrual products the cervix, but also in ensuring easy and rapid discharge and preventing infection and adhesions of the endometrium. The endometrium and cervix are sites of marked fibrinolytic activity. Plasminogen activators are demonstrated in the myometrium, endometrium and menstrual blood. The concentration of plasminogen activators in menstrual blood is maximal on the first day of bleeding, also higher in samples taken from the uterus than from vagina. It suggests that the activators are rapidly consumed and rarely clot forms in uterus than vagina.

   - Heparin like activity is also seen in uterine fluid, but it decreases at menstruation and increases in the menstrual cycle.

ABNORMAL UTERINE BLEEDING

- The most common reason for performing an endometrial biopsy is abnormal uterine bleeding, a term that refers to any nonphysiologic uterine bleeding. Age and menstrual/ menopausal status are especially important data, as causes of abnormal uterine bleeding vary significantly according to the age and menstrual status of the patient, as discussed later.

- Abnormal uterine bleeding is a common sign of a number of different uterine disorders ranging from dysfunctional (nonorganic) abnormalities or complications of pregnancy to organic lesions such as polyps, hyperplasia, or carcinoma. [15, 16, 17, 18, 19, 20, 21, 22] Several clinical terms are employed to describe different patterns of uterine bleeding.

- The prevalence of the various abnormalities that lead to abnormal bleeding is difficult to determine precisely, varying with the patient population and the terms used by investigators. [23, 24, 25] A practical approach to the possible diagnoses associated with abnormal bleeding takes age into account.

- Pregnancy- related and dysfunctional disorders are more common in younger patients whereas atrophy and organic lesions become more frequent in older individuals. [6] Polyps in perimenopausal and postmenopausal patients have been found in 2% to 24% of patients. Hyperplasia is found in up to 16% of postmenopausal patients undergoing biopsy, and endometrial carcinoma in fewer than 10% of patients. [26, 27] One consistent observation in studies of postmenopausal patients is the finding that atrophy is a common cause of abnormal bleeding, being found in 25% or more of cases. [26]

Clinical terms for abnormal uterine bleeding

Amenorrhea	Absence of menstruation
Hypermenorrhea	Uterine bleeding occurring at regular intervals but increased in amount. The period of flow is normal.
Hypomenorrhea	Uterine bleeding occurring at regular intervals but decreased in amount. The period of flow is the same or less than the usual duration.
Menorrhagia	Excessive uterine bleeding in both amount and duration of flow occurring at regular intervals
Metrorrhagia	Uterine bleeding, usually not heavy, occurring at irregular intervals
Menometrorrhagia	Excessive uterine bleeding, usually with prolonged period of flow, occurring at frequent and irregular intervals
Oligomenorrhea	Infrequent or scanty menstruation. Usually at intervals greater than 40 days
Abnormal uterine Bleeding (AUB)	A term that describes any bleeding from the uterus. Menorrhagia, Metrorrhagia, menometrorrhagia, and postmenopausal bleeding are all forms of AUB.
Dysfunctional uterine Bleeding (DUB)	Abnormal uterine bleeding with no organic cause. The term Implies bleeding caused by abnormalities in ovulation or follicle development and is a disorder of premenopausal women.
Postmenopausal	Abnormal uterine bleeding that occurs at least 1 year after menopause (the cessation of menses)

- Abnormal uterine bleeding can be categorized into two broad categories the first is due to organic causes and the second is caused by anovulation or oligo-ovulation (DUB).

1) **General systemic diseases** – acute pyrexial illness sometimes precipitate the onset of a period prematurely.
2) **Coagulation defects** - any blood disorders characterized by coagulation or by excessive capillary fragility can cause endometrial haemorrhage like thrombocytopenic purpura, aplastic anaemia, Leukemias, Vonwillebrand disease, anaemia and Christmas disease.

3) **Endocrine disorders; hyperoestrogenism**
- Hypothyroidism - tends to cause menorrhagia or polymenorrhoea.
- Hypothalamic and pituitary diseases cause excessive uterine bleeding or failure of normal cyclical pattern eg- acromegaly.
- Cirrhosis of liver and chronic renal disease disturbs the normal metabolism and inactivation of estrogen and its secretion can lead to menorrhagia or metrorrhagia.

Exogenous estrogen

- One of the commonest causes for metrorrhagia and menorrhagia. Administered by various routes for a variety of conditions such as pruritis vulvae; climacteric symptoms, for the control of uterine bleeding. Tamoxifen therapy for breast cancer is associated with irregular bleeding.

4) **Pelvic pathology:**
- Pregnancy states - like implantation bleeding, abortion and gestational trophoblastc disease can cause excessive uterine bleeding.
- Errors in uterine development - like uterus didelphys or bicornis is associated with an increased bleeding.
- Infections - all forms of pelvic peritonitis, salpingo-oopharitis and cellulitis tend to cause abnormal uterine haemorrhage. Chronic cervicitis with chlamydial infection can cause irregular bleeding especially contact bleeding.
- Local injury - foreign bodies - trauma to the interior of the uterus or IUCD retained in the cavity can result in either menorrhagia or intermittent or persistent acyclical bleeding and spotting.
- Displacements - abnormal uterine bleeding occurs in association with fixed retroversions, puerperal retro displacement, ovarian prolapse and uterovaginal prolapse.

- Endometriosis - with ovarian involvement it causes polymenorrhoea or polymenorrhagia is likely. On the other hand adenomyosis of uterus causes menorrhagia.

5) Tumours
- Ovarian - follicular cysts and neoplasms like oestrogen producing tumours cause acyclical anovular bleeding, non functional tumours can cause slight postmenopausal bleeding, large tumours can cause bleeding episodes as by disturbing the blood supply to uterus.
- Uterine
    - A) Leiomyomas cause progressive menorrhagia, polymenorrhoea and metrorrhagia depending upon the site.
    - B) Haemangiomas can also cause menorrhagia or metrorrhagia.
    - C) Malignant tumours - usually present as postmenopausal bleeding.
    - D) Surface growths - like endometrial polyp and other malignant growths encroaching on the uterine cavity, cause irregular or continuous bleeding, when ulcerated.
    - E) Chronic symmetrical enlargement of uterus. It is sometimes associated with excessive uterine bleeding, involving both myometrium and endometrium like - idiopathic development hypertrophy, prolonged and unopposed estrogen influence, local adhesion in the pelvis, sedentary occupation, varicocele in pampiniform plexus in broad ligament and etc.

6) Deep venous thrombosis - rare but very real cause of menorrhagia.

7) Psychological upsets - emotional and nervous disorders can cause excessive bleeding rather than amenorrhea. Like changes in environment, nervous tension, anxiety states, unsatisfied sex urge, marital upset, stress situation and redundancy or over work. These factors operate through the endocrine system influenced by hypothalamus or through the autonomic nervous system which controls the blood vessels supplying the pelvic organs.

Causes of abnormal uterine bleeding in adolescence

- Dysfunctional bleeding
  Anovulatory cycles
- Complications of pregnancy
- Endometritis
- Clotting disorders

Causes of abnormal uterine bleeding in the reproductive years

- Complications of pregnancy
- Endometritis
- Dysfunctional bleeding
    - Anovulatory cycles
    - Inadequate luteal phase
    - Irregular shedding
- Organic lesions
    - Leiomyomas
    - Polyps (endometrial, endocervical)
    - Adenomyosis
- Exogenous hormones
    - Birth control
    - Progestin therapy
- Hyperplasia
- Neoplasia
    - Endometrial carcinoma
    - Cervical carcinoma
- Clotting disorders

Causes of abnormal uterine bleeding in perimenopausal years

Dysfunctional bleeding

    Anovulatory cycles

Organic lesions

    Hyperplasia

    Polyps (endometrial, Endocervical)

Exogenous hormones

    Birth control

    Estrogen replacement

    Progestin therapy

Complications of pregnancy

Endometritis

Adenomyosis

Neoplasia

    Cervical carcinoma

    Endometrial carcinoma

    Sarcoma

Clotting disorders

Causes of abnormal uterine bleeding in postmenopausal years

Atrophy

Organic lesions

    Hyperplasia

    Polyps (endometrial)

Neoplasia

    Endometrial carcinoma

Exogenous hormones

    Estrogen replacement

    Progestin therapy (e.g., therapy of breast carcinoma)

Endometritis

Sarcoma

Clotting disorders

- Even among perimenopausal and postmenopausal patients, the proportion of cases attributable to any of the aforementioned conditions is age dependent. Atrophy and carcinoma occur more frequently in patients older than 60 years of age, while polyps and hyperplasia are more common in patients who are perimenopausal or more recently postmenopausal. In addition to these uterine causes of bleeding, other abnormalities, such as atrophic vaginitis, can cause vaginal bleeding, and this may be difficult to distinguish from uterine bleeding until the patient undergoes thorough clinical evaluation.

- Ideally, the clinical history that accompanies an endometrial sample should include some description of the pattern and amount of bleeding. Often in a patient of reproductive age or one who is perimenopausal the history is simply dysfunctional uterine bleeding (DUB). Clinically, this term suggests no other causes of bleeding except ovarian dysfunction. In this scenario the clinician performs the biopsy to rule out an organic lesion.

- A history of anovulation, obesity, hypertension, diabetes, and exogenous estrogen use should alert the pathologist that the patient is at increased risk for hyperplasia and adenocarcinoma, but this information is rarely included on the requisition. Typically, there is little accompanying clinical data except the patient's age and a short history of abnormal bleeding. Consequently, hyperplasia and adenocarcinoma must be diagnostic considerations for most endometrial specimens received in the laboratory. On rare occasions, hyperplasia or even adenocarcinoma is found in biopsies performed during an infertility workup where the clinical question was histologic dating rather than suspicion of these disorders. [28]

Infertility Biopsy

- When a patient undergoes biopsy for evaluation of infertility, the clinical information often is limited, but here, too, the history should include the date of the last menstrual period (LMP) to place an approximate time in the menstrual cycle. This information is useful but not precise for determining the actual day of the cycle, as ovulatory frequency and length of the follicular phase are highly variable among patients. Usually the main objective of biopsies for infertility is to determine whether there is morphologic evidence of ovulation and, if so, the histologic date. The gynaecologist may seek other specific information, such as response to hormone therapy, so it is important that the pathologist be given any additional history that may be necessary for the interpretation.

Products of Conception

- When endometrial sampling is performed to remove products of conception, clinical information often is sparse, as the main goal of the procedure is simply to remove the placental and fetal tissue. Significant pathologic changes are rare. Nonetheless, it is helpful to know if pregnancy is suspected, and, if so, the approximate gestational age of the pregnancy. If there is a suspicion of trophoblastic disease, this should be stated. In such instances the human chorionic gonadotropin (hcg) titer is relevant. If an ectopic pregnancy is suspected, alerting the pathologist can ensure rapid processing and interpretation of the specimen.

Hormone Therapy

- Because the endometrium is responsive to hormones, the history of hormone use is important information. Clinical uses of steroid hormones (estrogens, progestins, or both) include oral contraceptive use; postmenopausal replacement therapy; and therapy for endometriosis, hyperplasia, DUB, infertility, and breast carcinoma. As with other facets of the clinical data; this information may be absent or, if present, unreadable on the requisition. Consequently, the pathologist must be prepared to recognize hormonal effects in the absence of history indicating use of hormones.

Dysfunctional uterine bleeding [29]

- Dysfunctional abnormalities are frequent causes of uterine bleeding in perimenopausal and perimenarcheal women, and they occur to a lesser extent in women of reproductive age. DUB occurs either because of lack of ovulation following follicular development (anovulatory cycles) or because of luteal phase abnormalities. The latter include luteal phase defects (lpds) and abnormal persistence of the corpus luteum (irregular shedding). Often, before a biopsy is performed, DUB is managed by hormonal therapy. When bleeding persists, curettage often becomes necessary to control bleeding and exclude organic lesions.

- To place dysfunctional abnormalities in the appropriate pathophysiologic context, these disorders can be grouped into two broad categories: estrogen-related and progesterone-related bleeding. The more common is estrogen related DUB, which refers to episodes of bleeding that are related to lack of ovulation with alterations in endogenous estrogen levels. Although not really a manifestation of DUB, atrophy is included as a form of estrogen related bleeding because it occurs when the endometrium is deprived of estrogen for a relatively long period of time. In clinical classifications of bleeding disorders, atrophy is not regarded as a form of dysfunctional uterine bleeding, yet it is a significant cause of abnormal bleeding. The second, less frequent category of DUB is progesterone related and reflects abnormal endogenous progesterone levels.

- All these disorders, classified here as DUB, reflect variations in ovarian hormone production. Exogenous hormones may produce endometrial patterns that are indistinguishable from the patterns seen in DUB caused by endogenous hormone fluctuations.

Estrogen-Related Bleeding

Proliferative with Glandular and Stromal Breakdown

- This term describes the endometrial changes resulting from anovulatory cycles. It is probably the most common abnormality found in biopsies performed for abnormal bleeding in perimenopausal women. Anovulatory cycles with bleeding also occur in perimenarcheal adolescents in whom regular ovulatory cycles are not established. Anovulatory bleeding even occurs sporadically in women throughout the reproductive years. Usually, this bleeding does not lead to the need for biopsy in younger patients, as the risk of other lesions, especially hyperplasia and carcinoma, is low in individuals of this age. An exception is women with chronic anovulation associated with the Stein– Leventhal syndrome (polycystic ovarian disease), because these women have an increased risk of development of hyperplasia or carcinoma.

- Anovulatory cycles result when a cohort of ovarian follicles begins to develop but ovulation does not occur. Chronic anovulation may be the result of a variety of disorders, including hypothalamic dysfunction and obesity, because of peripheral conversion of androgens to estrogen in adipose tissue, as well as increased androgen production by the adrenal glands or the ovaries. Causes of anovulation following recruitment of follicles are complex. They include defects in the hypothalamic–pituitary– ovarian axis such as hyperprolactinemia, abnormal feedback mechanisms of hormonal control, and local ovarian factors that interfere with appropriate follicular development. Whatever the pathogenesis, if ovulation does not occur, a corpus luteum does not develop, and progesterone is not produced. The follicles produce estradiol, which stimulates endometrial growth. The developing follicles may persist for variable periods of time before undergoing atresia. As long as the follicles persist, estradiol is produced and the endometrium proliferates.

- When these follicles undergo atresia, estradiol production falls precipitously and estrogen withdrawal bleeding occurs. In this instance, the loss of estrogenic support of endometrial proliferation results in destabilization of lysosome membranes and vasoconstriction with bleeding. In contrast to estrogen withdrawal bleeding, estrogen breakthrough bleeding results from persisting follicles that produce estradiol; the proliferating endometrium becomes thicker and outgrows its structural support. Focal vasoconstriction and thrombosis of dilated capillaries it. In either event, the result is irregular breakdown and bleeding of the endometrium. In most cases of anovulatory DUB, the endometrium shows a proliferative phase pattern with glandular and stromal breakdown. The amount of tissue and the architectural pattern of the glands depend on the duration of unopposed estrogenic stimulation, not necessarily the level of estrogen.

- Chronic anovulation results in persistence of follicles and sustained unopposed estrogen stimulation. A greater quantity of endometrial tissue develops with actively proliferating glands and augmented glandular tortuosity. Dilated venules appear in the subepithelial stroma and often thrombose. Because of continuous estrogenic stimulation, the tissue often shows estrogen induced epithelial changes ("metaplasia"), especially ciliated cell and eosinophilic cell change. The glands also may show focal subnuclear vacuolization as a response to estrogen stimulation, but the extent and uniformity of the vacuolization are less than that seen in normal early secretory glands. The cytoplasmic changes and subnuclear vacuoles complicate the interpretation of the histologic pattern but do not change the diagnosis. Prolonged, unopposed estrogenic stimulation also can lead to the development of varying degrees of hyperplasia, atypical hyperplasia, and even well differentiated adenocarcinoma, but these organic lesions are not functional disorders and, as such, are not considered causes of DUB.

- When proliferative endometrium shows breakdown and bleeding, the pattern strongly suggests anovulatory cycles. Exogenous estrogens can cause similar patterns, and therefore a complete clinical history is needed to be certain that the bleeding pattern is truly due to anovulation. The differential diagnosis of proliferative phase endometrium with glandular and stromal breakdown also includes inflammation, polyps, and leiomyomas. In such cases, the presence of other features, such as plasma cells in chronic endometritis or the dense stroma and thick-walled vessels of polyps, establishes the proper diagnosis.

Disordered Proliferative Phase and Persistent Proliferative Phase

- When chronic anovulatory cycles result in abundant proliferative tissue, mild degrees of disorganization characterized by focal glandular dilation may occur. Usually these are regarded as variants of normal proliferative endometrium. Sometimes more sustained estrogen stimulation may result in the focal branching and some dilation of glands, yielding a proliferative phase pattern that is neither normal nor hyperplastic. The terms "disordered proliferative phase pattern" and "persistent proliferative phase" have been applied to describe this pattern of proliferative endometrium with tortuous and mildly disorganized glands. A designation of a "disordered" or "persistent proliferative phase" has utility in correlating the morphologic findings with the apparent pathophysiology.

- Term "disordered proliferative" often is inappropriately applied to a variety of patterns, including normal proliferative endometrium, proliferative endometrium with breakdown, artifactual crowding, basalis, and simple hyperplasia. The diagnosis of disordered proliferative phase should be reserved for cases in which assessment is based on intact, well oriented fragments of tissue. In these areas the abnormal glands should be focal. These glands are qualitatively similar to those seen in simple hyperplasia, but they are limited in extent and interspersed among glands with a normal proliferative phase pattern. This criterion helps to separate the focal disordered proliferative phase pattern from simple hyperplasia, a more diffuse abnormality. If the tissue is extensively fragmented or disrupted by the procedure and contains mainly detached proliferative glands, it is best to diagnose the change only as proliferative. Extensive breakdown in proliferative endometrium can also display a disorganized appearance to the glands because of fragmentation, but again this change is not that of a true disordered proliferative phase pattern.

Atrophy

- Atrophy is an important cause of abnormal uterine bleeding in postmenopausal patients, found in 25% or more of cases coming to biopsy. In many laboratories, atrophy is found in up to 50% of biopsy specimens taken for postmenopausal bleeding, and in one study 82% of cases of postmenopausal bleeding were attributable to atrophy. [30]

- Besides being common in postmenopausal patients, atrophic endometrium can occur in reproductive-age patients with premature ovarian failure, either idiopathic or due to radiation or chemotherapy for malignancies.

- With atrophy, tissue obtained at biopsy is typically scant, often consisting only of a small amount of mucoid material. Characteristically, atrophic endometrium is composed of tiny strips and wisps of surface endometrium and detached, fragmented endometrial glands. The epithelium is low columnar to cuboidal with small, dark nuclei and minimal cytoplasm. Stroma is scant or absent, consisting of a few clusters of small spindle cells. Mitotic activity is absent. The cystic change seen in atrophic glands in hysterectomy specimens is not observed in biopsies because tissue fragmentation from the procedure disrupts the glands. Breakdown and bleeding may be superimposed on the features of atrophy, although often, even when there is a history of abnormal uterine bleeding, the sections show no evidence of glandular and stromal breakdown.

- Although there is a paucity of tissue in biopsy specimens of atrophic endometrium, these specimens are not insufficient or inadequate. The scant tissue may be all that is present and therefore is completely representative of the lining of the uterine cavity. The minimal quantity of tissue should serve as a clue to the diagnosis; it does not represent an insufficient specimen.

Progesterone-Related Bleeding

- Biopsy specimens from reproductive-age and perimenopausal women occasionally show abnormal secretory phase patterns with associated nonmenstrual breakdown and bleeding. In such cases the pattern is secretory owing to ovarian progesterone production, but the glandular and stromal changes usually are less advanced than those seen in normal late secretory endometrium. The endometrial pattern does not correlate with any date of the normal luteal phase. The glands may show secretory changes yet lack marked tortuosity and secretory exhaustion, while the stroma lacks extensive predecidual change. In other cases the glands appear to show a "hypersecretory" pattern, with vacuolated cytoplasm, marked tortuosity, and luminal secretion, while the stroma lacks predecidual change. In addition, the tissue shows foci of breakdown with characteristic changes such as nuclear dust, fibrin thrombi, and dense cell clusters, similar to that which occurs in the proliferative endometrium with glandular and stromal breakdown. Often in abnormal secretory bleeding patterns, the glands show stellate shapes as they involute. This latter pattern of collapsing, star-shaped secretory glands is nonspecific, however, and simply shows secretory gland regression that could be due to a variety of factors.

- These changes may reflect DUB due to luteal phase abnormalities that include lpds and irregular shedding. The etiology and frequency of dysfunctional bleeding caused by luteal phase abnormalities are not known, however, as these disorders appear to be sporadic and do not persist long enough to permit clinical–pathologic correlations.

Luteal Phase Defects

- Of the two defined luteal phase abnormalities that may cause abnormal bleeding, LPD probably occurs more frequently. The corpus luteum is "insufficient" in LPD, either regressing prematurely or failing to produce an adequate amount of progesterone to sustain normal secretory phase development. This is a sporadic disorder of the reproductive and perimenopausal years. With LPD, ovulation occurs, so secretory changes develop. If abnormal bleeding is the result, the appearance is that of breakdown and bleeding in a nonmenstrual secretory phase pattern. Glands with secretory changes, including basally oriented nuclei and vacuolated cytoplasm, but lacking the tortuosity of late secretory phase glands, characterize the pattern. Focal breakdown is present with "stromal blue balls" and karyorrhectic debris. This pattern is nonspecific.

Irregular Shedding

- Irregular shedding is attributed to a persistent corpus luteum with prolonged progesterone production. This is the least studied and consequently the most poorly understood form of dysfunctional patterns. One pattern of irregular shedding yields a mixed phase pattern composed of secretory and proliferative endometrium. The diagnosis is reserved for those specimens in which there is a mixed pattern of secretory and proliferative glands at least 5 days after the onset of bleeding. Irregular shedding is also manifested by irregular secretory phase development in which different foci show more than 4 days' difference in the morphologic date. Breakdown and bleeding with glandular and stromal collapse is present, usually focally, but occasionally in a diffuse pattern. Frequency of irregular shedding as a cause of DUB is not known.

Abnormal secretory Endometrium with Breakdown of Unknown Etiology

- A variety of other factors may also be associated with a pattern of aberrant secretory phase development with superimposed bleeding .For example, endometrial changes associated with abortions or ectopic pregnancies, response to exogenous progestins, tissue near a polyp, endometrium overlying leiomyomas, and endometrium involved with inflammation or adhesions all can show patterns of abnormal secretory development and bleeding. Other poorly understood ovarian disorders, such as a luteinized unruptured follicle, presumably could result in abnormal secretory endometrial changes. With this latter entity, developing follicles are believed to undergo luteinization of the granulosa and theca cells with progesterone production in the absence of ovulation. In addition to these considerations, management of DUB often involves progestational therapy. If bleeding is not controlled, curettage is performed. Accordingly, progestin effects superimposed on the underlying abnormality may complicate the histology. These patterns show glands with secretory changes such as basally oriented nuclei and diffuse cytoplasmic vacuolization and absence of mitotic activity. The glands may be tortuous. Often the stroma is dense, lacking edema or predecidua.The endometrium in such cases cannot be assigned to any histologic day of the normal secretory phase of the menstrual cycle. The changes of glandular and stromal breakdown are similar to those found in any bleeding phase endometrium with glandular and stromal collapse, "stromal blue balls," and eosinophilic syncytial change. With early breakdown, tortuous secretory glands often show star-shaped outlines.

Possible causes of nonmensrual secretory phase bleeding

Luteal phase defect

Persistent corpus luteum (irregular shedding)

Organic lesions

    Submucosal leiomyomas

    Intrauterine adhesions

    Inflammation

Complications of pregnancy

Progestin effects

ENDOMETRITIS

- Endometritis usually is a disorder of the reproductive years, although it may occur in postmenopausal patients. Endometrial inflammation typically accompanies pelvic inflammatory disease of the upper genital tract. [31, 32] Endometritis typically presents with intermenstrual vaginal bleeding, and sometimes it causes menorrhagia. In one study, 8 % of outpatient endometrial biopsies, most of which were done for abnormal bleeding, showed chronic endometritis. [33]

- Endometrial inflammation often is nonspecific and rarely has morphologic features that indicate a definite etiology. The nonspecific forms of endometritis traditionally have been separated into chronic and acute forms, depending on the type of inflammatory infiltrate; most are referred to as chronic nonspecific endometritis.

Nonspecific Endometritis

- Endometritis may be diffuse or focal and can range from a subtle inflammatory infiltrate to a pronounced inflammatory reaction. Endometritis typically shows a pattern of a mixed inflammatory infiltrate containing plasma cells and lymphocytes, and, not infrequently, neutrophils and eosinophils. In addition to inflammatory cells, there is a constellation of histologic findings that facilitate recognition of endometrial inflammation. The other morphologic changes include reactive stroma, epithelial changes, abnormal glandular development, and evidence of glandular and stromal breakdown. [33]

Inflammatory cells

- Plasma cells are the most important histologic feature for the diagnosis of endometritis. [34, 35, 36, 37] Their presence is required to establish the diagnosis of chronic endometritis, because, in contrast to lymphocytes, they are not present in normal endometrium. Plasma cells generally are most numerous in the periglandular and subepithelial stroma and around lymphoid aggregates. In cases in which the plasma cell infiltrate appears subtle or equivocal, the background pattern is as important as the quantity of plasma cells for establishing the diagnosis of endometritis.

- Normal endometrial stromal cells, especially predecidualized cells in the late secretory phase, can resemble plasma cells, having eccentric nuclei and a pale perinuclear zone. The plasma cell, however, is identified by its distinctive, clumped chromatin arrangement yielding a clock-face pattern. A methyl green pyronin histochemical stain, immunohistochemistry for immunoglobulin G or syndecan, and in situ hybridization for kappa and lambda immunoglobulin light chains can help demonstrate plasma cells when the cytologic features are not diagnostic by routine histology. Often the inflammatory infiltrate includes numerous lymphocytes that tend to concentrate in the subepithelial stroma. Lymphoid follicles become prominent and may show germinal centers; larger transformed lymphocytes and immunoblasts also may be interspersed.

- Neutrophils as a part of the inflammatory infiltrate indicate an acute process. This neutrophilic inflammatory infiltrate typically infiltrates the surface epithelium and extends into gland lumen, sometimes forming microabscesses in the glands. [33] Neutrophils, however, also can be present in menstrual endometrium, where foci of glandular and stromal breakdown are present, too, without signifying an infectious process. Therefore, like the presence of lymphocytes, the presence of neutrophils alone is not sufficient to indicate inflammation. The pattern of distribution of these cells in the endometrium and the accompanying cellular infiltrate, usually including plasma cells, must be considered before making the diagnosis of endometritis.

- Acute endometritis without a chronic (plasma cell) component is extremely unusual and occurs most frequently in the postpartum or postabortal patient. When acute endometritis is present, there is a neutrophilic infiltrate in the glands with microabscess formation and infiltration of neutrophils into the surface epithelium. Marked inflammation also will result in formation of granulation tissue with a network of small vessels in a fibroblastic stroma.

- On occasion eosinophils may be present as a part of the inflammatory infiltrate. [33] They are not normally present in the endometrium. Like lymphocytes or neutrophils, eosinophils should be present in a background of inflammatory changes to be a component of endometritis. Eosinophilic infiltrates also can occur following curettage, apparently as a result of the instrumentation, and in this case they represent a nonspecific response to the procedure.

Stromal Changes

- With endometritis the stroma typically shows reactive changes. Stromal cells become spindle-shaped, resembling fibroblasts, and are elongate and bipolar, in contrast to the rounded, ovoid shape of the nonreactive stromal cell. The reactive process is also characterized by a swirling, interlacing pattern of the spindle cells, which may from radial, "pinwheel" arrangements. Plasma cells usually are interspersed in the reactive stroma. Superficial stroma may become edematous. [33]

Abnormal Glandular Devlopment

- In cycling patients the endometrial response to hormones is often diminished. Usually the endometrium has proliferative phase characteristics, with tubular glands showing mitotic activity. In the secretory phase the glands may lose their normal pattern of reactivity. Secretory changes occur in ovulating women, but they often show abnormal development with less gland tortuosity and distension than is seen in a normal, noninflamed secretory phase. The changes can include irregular or retarded maturation of secretory phase endometrium. Glands may appear underdeveloped, lacking tortuosity and lumenal secretions. Plasma cells may rarely be seen in histologically normal secretory phase endometrium.

Epithelial changes

- Reactive cellular changes also affect the endometrial surface and glandular epithelium. The epithelium may show squamous and eosinophilic cell change, especially when the inflammation is long standing and intense. [33] The reactive epithelial cells may become stratified, with prominent nucleoli, cleared chromatin, and increased mitotic activity.

Glandular and Stromal Breakdown

- Endometritis also results in focal glandular and stromal breakdown. With severe and prolonged chronic inflammation; the changes of irregular bleeding with breakdown and regeneration become prominent. This pattern of haphazard bleeding leads to a corrugated surface with foci of regenerating and shedding endometrium interspersed. The inflamed stroma becomes dense and less responsive to hormonal changes. Because of the irregular growth, the tissue may become polypoid, and the resemblance to a polyp is accentuated by the dense stroma that accompanies the inflammation. A mixture of acute and chronic bleeding patterns also may be present, with areas of stromal collapse, glandular breakdown, stromal fibrosis, macrophages, and hemosiderin deposition.

➢ Specific Infections

- The inflammatory response associated with Chlamydia trachomatis infection is usually marked. The inflammatory infiltrate tends to be diffuse, with plasma cells, lymphocytes, and lymphoid follicles with transformed lymphocytes; Stromal necrosis and reactive atypia of the epithelium also can be present. These marked inflammatory changes are not specific for chlamydia, however, but appear to reflect the presence of upper genital tract infection and acute salpingitis. Chlamydia trachomatis or Neisseria gonorrheae are most frequently associated with these findings of a marked acute and chronic inflammatory infiltrate, although infections with C. Trachomatis produce a greater concentration of plasma cells and more lymphoid follicles than N. Gonorrheae. [28, 38]

➢ Granulomatous Inflammation

- Granulomatous inflammation of the endometrium is infrequent. Often the process is caused by mycobacterium, especially Mycobacterium tuberculosis, and the infection usually indicates advanced disease.

Tuberculous endometritis is more commonly encountered in endometrial biopsies undertaken during assessment of primary or secondary female infertility, as endometrial involvement is a reflection of more widespread disease that also affects the fallopian tubes in most cases. Tuberculous endometritis also can cause abnormal uterine bleeding in postmenopausal patients. In tuberculous infection the granulomatous response is variable. Often the granulomas are nonnecrotizing. Well-formed granulomas may be difficult to identify unless the endometrium is biopsied in the late secretory phase when the granulomas have had sufficient time to develop. The surrounding stroma can show a lymphocytic infiltrate. As with any form of inflammation, gland development may be altered, lacking an appropriate secretory response if the biopsy is taken in the luteal phase. Acid-fast stains rarely demonstrate the characteristic organism in endometrial infections, and culture of fresh tissue or PCR of paraffin-embedded tissue may be needed to establish the diagnosis. Fungal infections, including cryptococcosis, coccidioidomycosis, and blastomycosis, rarely involve the endometrium, resulting in granulomatous inflammation. [39, 40, 41]

➤ **Actinomycosis**

- Infection by Actinomyces israelii is another rare cause of endometritis. This organism typically is found in endometritis associated with use of the intrauterine contraceptive device (IUCD). When actinomycosis associated inflammation is present, the inflammatory response usually is intense, with many plasma cells, lymphocytes, and neutrophils present throughout the tissue. The organisms show the typical sulfur granule morphology and can be stained by Gram and methenamine-silver stains. [36]

➤ **Cytomegalovirus**

- Rarely, endometrial biopsy will show evidence of cytomegalovirus infection. This may occur in immunosuppressed patients or it may be found in women with no known underlying disorder. Regardless of immunologic status, the tissue shows the characteristic nuclear and cytoplasmic inclusions in epithelial cells and occasional endothelial cells. The stroma may show a sparse plasma cell infiltrate.

➢ **Herpesvirus**

- Herpesvirus rarely infects the endometrium, but it may occur, usually as an ascending process associated with cervical infection. When present in the endometrium, it can cause patchy necrosis of the glands and stroma. The diagnosis is established by identifying cells that show typical herpesvirus cytopathic effect. Cowdry type A inclusion and multinucleate cells with molded ground-glass nuclei can be found in the glandular epithelium or the stroma in areas of necrosis.

ENDOMETRIAL POLYPS

- Endometrial polyps are benign, localized overgrowths of endometrial glands and stroma that are covered by epithelium and project above the adjacent surface epithelium. Polyps arise as monoclonal overgrowths of genetically altered endometrial stromal cells with secondary induction of polyclonal benign glands through as yet undefined stromal – epithelial interactive mechanisms. Chromosomal analysis of polyp stroma shows in the majority of cases clonal translocations, involving 6p21-22, 12q13-15 or 7q22 regions.

- Endometrial polyps variably respond to circulating estrogens and progesterone. The glands in the polyp may sometimes be indistinguishable from those in the rest of the endometrium but this is unusual. If the stroma is made up predominantly of smooth muscle but the glands are not atypical, the term adenomyoma is employed. [42]

- Polyps occur over a wide age range, but are most common in women in the fourth and fifth decades, becoming less frequent after age 60. [37, 46] Usually they present with abnormal uterine bleeding, and have been implicated as a cause of abnormal bleeding in between 2 % and 23 % of patients coming to biopsy. [46] They also have been implicated as a possible cause of infertility, either by physically interfering with blastocyst implantation or by altering the development of secretory phase endometrium, making it less receptive to the implanting embryo. [47] Tamoxifen therapy is a risk factor for endometrial polyps, [48] and polyps are the most frequent pathologic lesion found in patients receiving tamoxifen therapy for breast carcinoma. [49] Tamoxifen related polyps may be multiple and large. Hormone replacement therapy is not a risk factor for polyps. [48] Large polyps that extend into the endocervix and dilate the internal os can cause endometritis.

MORPHOLOGICAL FEATURES:

- Endometrial polyps range in size from a slightly rounded protuberance to a large broad based or pedunculated, oval structure filling the uterine cavity. Many polyps are sessile and have a broad base of attachment. 20 % polyps are multiple. The surface may be smooth and shiny, often hemorrhagic particularly at the tip. The cut surface may be uniform or it may show cysts, hemorrhage and necrosis.

- There are five histo-morphological forms of common benign endometrial polyps: proliferative/hyperplastic, atrophic, functional, mixed endometrial–endocervical, and adenomyomatous. The atypical polypoid adenomyoma represents a distinctive form of polyp that should be segregated from the variants of common polyps.

- Microscopy: The diagnosis of a polyp in a curetting depends upon the finding of at least two of three particular histologic features and exclusion of mimics. These are (i) irregularly shaped and positioned glands (ii) stroma altered by fibrosis or excessive collagen (iii) thick walled blood vessels.

- The prevalence of polyps in general population is about 24 %. Polyps are common in women over 40 years of age and extremely rare before menarche. The most common presentation is postmenopausal bleeding in older patients and intermenstrual bleeding in younger patients. A polyp should always be considered if abnormal bleeding persists after curettage because polyps that contain a delicate, pliable stalk may elude the curette. Polyps are believed commonly to be a risk factor for endometrial cancer because hyperplastic and neoplastic lesions can be found in their context.

- Most of the published reports are based on specimens that were obtained by curettage or by other sampling techniques such as pipette. In studies, fragmentation of the polyp, incompleteness of the specimen and the presence of adjacent endometrium hamper the diagnosis as to wheather endometrial cancer has originated in an antecedent benign polyp or in the surrounding endometrium. [37, 43]

- Glands in polyps often fail to cycle normally, secretory changes may be weak or absent in contrast to the surrounding endometrium or the glands may appear dilated and inactive. Squamous metaplasia may be present. The mesenchymal component of polyps may consist of endometrial stroma, fibrous tissue, as smooth muscle but generally the stroma appears more fibrous than normal fundic endometrium. Polyps are morphologically diverse lesions that are difficult to sub classify; but most can be categorized as hyperplastic, atrophic or functional. Hyperplastic polyps contain proliferating, irregularly shaped glands resembling diffuse, nonpolypoid endometrial hyperplasia probably etiologically related to hormone imbalances. Atrophic polyps consist of low columnar or cuboidal cells lining cystically dilated glands. These polyps are typically found in postmenopausal patients and may represent regression of hyperplastic or functional polyps. Functional polyps containing glands resembling normally cycling endometrium are relatively uncommon.

- Polyps may be difficult to recognize in curettage specimens, ideally they appear as polypoid shaped fragments of tissue with epithelium on three sides. These criteria may be difficult to apply if lesions are fragmented or partially removed. In addition, normal endometrium has an irregular surface that may appear as a polypoid fragment with epithelium on three sides when sectioned tangentially. Identification of tissue fragments containing irregular glands dense or fibrous stroma or thick walled vessels that contrast with the appearance of the surrounding endometrium suggest a polyp. [37]

- **Robboy et al.**, studied 151 endometrial polyps and described a new feature useful in the diagnosis of polyps, namely endometrial glands that grow parallel to each other with their long axis parallel to the elongate sides of a pedunculated polyp. Endometrial glands in their normal physiological stage grow with their long axis perpendicular to the mucosal surface that lines the endometrial cavity. The parallel arrangement was more prominent in pedunculated polyps than in sessile polyps, in premenopausal than in postmenopausal women and in functional and hyperplastic polyps than in fibrous polyps. [44]

- **Savelli L. Et al.**, studied 358 endometrial polyps and divided them into group A: benign and group B: atypical hyperplastic and cancerous. Age, menopause status, and hypertension were associated significantly with group B. [43]

- **Kelly P. Et al.**, studied 1031 endometrial polyps and found that 3.1 % cases were having hyperplasia occurring in polyps whereas other studies have found the incidence to be between 11 and 29 %. The authors claim that the term hyperplastic polyps are often used loosely to denote proliferative activity rather than true hyperplasia within a polyp. [45]

Atypical Polypoid Adenomyoma

- This is an unusual and distinctive polyp characterized by glands that are lined by atypical epithelium and surrounded by cellular smooth muscle and variable amounts of fibrous tissue. The atypical polypoid adenomyoma typically occurs in premenopausal or perimenopausal women, with a mean age of about 40 years. Premenopausal women often are nulliparous. [50]

- The glands of the atypical polypoid adenomyoma are haphazardly arranged but generally are not markedly crowded or back-to-back. They resemble the glands in simple atypical hyperplasia. The glandular cells have enlarged, stratified, and rounded nuclei with a vesicular chromatin pattern and prominent nucleoli. The cytoplasm is eosinophilic and the glands resemble those found in atypical hyperplasia. A very characteristic though not specific feature is the presence of squamous change (metaplasia), containing central nonkeratinizing nests of squamous cells. Central necrosis may occur in the squamous nests.

- Smooth muscle encompasses the glands, and endometrial stroma is largely absent. The smooth muscle is arranged in short interlacing fascicles that contrast with the elongate bundles of smooth muscle found in normal myometrium or in adenomyomatous polyps. The smooth muscle component can show increased mitotic activity, with up to two mitoses per 10 high-power fields, but there is no evidence of cytologic atypia.

Effects of Hormones

- Women receive hormone preparations for a variety of reasons, including birth control and treatment for dysfunctional uterine bleeding, perimenopausal and postmenopausal symptoms, endometriosis, endometrial hyperplasia and carcinoma, breast carcinoma, and certain types of infertility. Usually the exogenous hormone is some form of progestin, but estrogenic and even androgenic hormones are used for some disorders. The endometrium shows the effects of these hormones.

- An endometrial biopsy or curettage may be performed when abnormal bleeding occurs or when hormone therapy does not correct abnormal bleeding that is thought to be dysfunctional. Sometimes, however, the biopsy is intended to evaluate the status of the endometrium following hormonal therapy, as in the case of hyperplasia managed with progestin therapy or routine follow-up of patients on hormone replacement therapy. In other circumstances the endometrial sampling is coincidental with another procedure, such as tubal ligation, where the patient has received hormone therapy to ensure no interval pregnancy.

- The effects of different types of hormones on the endometrium: (1) hormones used in women of reproductive age that clearly have estrogenic or progestogenic effects, such as oral contraceptives; (2) estrogen–progestin hormone replacement therapy in postmenopausal women; (3) tamoxifen therapy for breast cancer; and (4) other hormones with less well established effects on the endometrium.

- **Estrogenic hormones**

- Estrogen therapy is largely used in perimenopausal or postmenopausal women to treat symptoms of the menopause, such as vasomotor instability, atrophic vaginitis, and osteoporosis. [51] Use of estrogenic hormones by themselves is associated with an increased risk of developing endometrial adenocarcinoma, so the use of these hormones alone is now unusual in patients with a uterus. Consequently, the effects of unopposed exogenous estrogen are seen less frequently in biopsy specimens than the effects of combined estrogen– progestin compounds, as progestins abrogate the effect of estrogen stimulation on the uterus.

- Unopposed estrogenic stimulation causes the endometrium to proliferate. The result is variable, depending on the dose and duration of use. Often the pattern is that of proliferative phase endometrium, showing tubular to tortuous glands and abundant stroma. The patterns can be identical to those seen with anovulatory cycles and may include superimposed breakdown and bleeding. Continued, prolonged estrogenic stimulation can lead to disordered proliferative phase patterns and hyperplasia. Estrogen-related epithelial cytoplasmic changes, especially squamous differentiation and ciliated cell change, also often occur.

- In some patients continued estrogen use leads to atypical hyperplasia and adenocarcinoma. [52] The risk of malignancy increases with the duration of therapy. Usually unopposed estrogen use for at least 2 to 3 years is found in patients who develop adenocarcinoma, and the highest risk is in patients who have taken estrogens for 10 years or longer. The duration of therapy generally is more important than the dose of the estrogen.

- **<u>Progestins and Oral Contraceptives</u>** [52]

  - Progestin effects are common; the subject of progestin-related changes is complex. Various forms of these synthetic analogues of progesterone, also termed "progestogens" or "progestagens," are widely used, either alone or in combination with an estrogen. Progestinonly therapy is useful in the empirical medical management of abnormal uterine bleeding that clinically appears to be dysfunctional. These hormones, such as medroxyprogesterone acetate or norethindrone acetate, suppress ovulation and endometrial growth. They also lead to secretory maturation and progesterone withdrawal bleeding.

  - Some progestins, such as oral or injectable medroxyprogesterone acetate, may also be used to treat neoplasia of the breast or endometrium. Still others, such as norgestrel, are used for contraception. Progestins, especially oral contraceptives, also are used to treat endometriosis.

  - The morphologic appearance of the endometrium following progestin therapy is variable and depends on the underlying status of the endometrium as well as the dose, potency, and duration of progestin therapy.

- The effects of the progestins can be placed into three general morphologic patterns that form the basis for understanding the entire spectrum of progestin- mediated changes. These patterns include: (1) decidual (pregnancy-like) changes, (2) secretory changes, and (3) atrophic changes.

Morphologic features of progestin effects

Decidual (pregnancy-like) effects

    Abundant tissue, often polypoid

    Glands show marked secretory activity

    Stroma appears decidualized with lymphoid infiltrate

    Vascular ectasia

Secretory effects

    Moderate to sparse amount of tissue

    Mildly tortuous secretory glands lined by columnar cells.

    Stromal cells plump, oval (predecidual)

    Vascular ectasia

Atrophic effects

    Sparse tissue

    Glands small and atrophic, not coiled

    Variable amount of stroma with plump to spindle shaped Cells.

- **Tamoxifen** [53]

  - Tamoxifen is a nonsteroidal antiestrogen that is widely used in the hormonal therapy of breast carcinoma. This drug is a selective estrogen receptor modulator (SERM) with its action mediated through the estrogen receptor. The effect of tamoxifen on the endometrium appears to depend on the menopausal status and the dose and duration of tamoxifen use. Both endometrial hyperplasia and carcinoma occasionally occur in patients on tamoxifen, and some studies suggest an increased incidence of both these disorders in patients receiving tamoxifen.

- Carcinosarcomas and other sarcomas including endometrial stromal sarcoma and adenosarcoma also have been reported in patients on tamoxifen. Endometrial polyps appear to be one of the most common pathologic findings in patients on tamoxifen. These patients are postmenopausal and have received long-term tamoxifen therapy for metastatic breast carcinoma. The polyps tend to be large and multiple and they may be recurrent. The stroma is variably edematous, myxoid, or fibrous. Often these polyps show mildly hyperplastic changes. Various types of cytoplasmic change or metaplasia have been described in the glands, especially mucinous and clear cell change. Occasionally endometrial polyps in patients receiving tamoxifen show foci of secretory changes in the glands with clear to vacuolated cytoplasm.

- **Clomiphene Citrate**

  - Clomiphene citrate is another antiestrogen that is used to induce ovulation in the treatment of infertile patients who are anovulatory. [51] It also may be used to treat luteal phase defects (lpds). This hormone stimulates multiple follicles to develop, and ovulation follows. It is thought to act by competitively binding to estrogen receptors in the hypothalamus, causing increased levels of follicle-stimulating hormone (FSH) and luteinizing hormone (LH) that induce ovulation. Like tamoxifen, clomiphene citrate has been found to have estrogenic as well as antiestrogenic activity.

  - Effects of clomiphene citrate on the morphologic pattern are that of normally developing secretory phase endometrium, which can be histologically dated, and this has been our experience. Often the histologic date correlates with the chronological postovulatory date, but sometimes the endometrium shows a significant lag in development. It is postulated that clomiphene citrate may cause lpds.

- **Danazol**

  - Danazol is structurally related to testosterone and is a weak androgen. Its main metabolite, ethisterone, is a weak progestin, however. This steroid is used for the treatment of endometriosis. Because it suppresses endometrial growth, it also may be used to treat menorrhagia and endometrial hyperplasia. [51] Within a few months of use, the amount of tissue is reduced. Glands show weak and irregular secretory changes with mild tortuosity, basal nuclei, and some cytoplasmic vacuolization. The stroma is hypercellular. With prolonged therapy, the glands show atrophy with scant to no secretory activity. Vascular ectasia also can occur. Occasional patients will show some proliferative activity with stromal and glandular mitoses.

- **Antiprogestin RU 486**

  - The synthetic progestogenic steroid RU 486, or mifepristone, has high affinity for progesterone receptors in the endometrium, which causes its antiprogesterone action. Its main use is for spontaneous termination of early pregnancy. The drug is used in high dose to induce early abortion. This drug also has been assessed for possible contraceptive use at much lower doses, and preliminary studies suggest that with longer term use it retards the secretory phase.

  - Low-dose treatment inhibits ovulation and menstruation. Thus, mefepristone results in inhibition of glandular secretory activity with degenerative and vascular changes. Several reports have indicated that the glands may become irregular in size and shape, with dilated glands. The stromal response appears to be variable, but some reports indicate that the stroma remains dense with increased mitotic activity. [54]

Endometrial Hyperplasia

- The diagnosis and management of endometrial hyperplasia have been unnecessarily complicated by the use of a wide variety of terms and histologic classifications. Terms such as "adenomatous hyperplasia," "atypical hyperplasia," and "carcinoma in situ" have been used by different authors for the same lesions, and, conversely, different investigators have used the same term to describe different lesions. The distinction of atypical hyperplasia from well-differentiated adenocarcinoma has been further clouded by the term "carcinoma in situ." The confusion resulting from the use of different classifications often precluded comparison of data between institutions and created problems in communication between the gynecologist and the pathologist. The World Health Organization (WHO) and the International Society of Gynecologic Pathologists (ISGYP) have promoted one classification of endometrial hyperplasia that has gained widespread acceptance. [55]

World Health Organization classification of endometrial hyperplasia

Hyperplasia (without atypia)

   Simple

   Complex

Atypical hyperplasia

   Simple

   Complex

- All types of hyperplasia are characterized by an increase in the gland-to-stroma ratio, irregularities in gland shape, and variation in gland size. The amount of stroma separating the glands distinguishes simple and complex forms of hyperplasia, regardless of the presence of atypia.

Hyperplasia without Atypia

Morphologic features of endometrial hyperplasia without atypia

Cytologic features

   Nuclei

      Pseudostratified

      Cigar-shaped to oval with smooth contours

      Uniform chromatin distribution

      Small to indistinct nucleoli

      Mitotic activity, variable amount

   Cytoplasm

      Variable, often amphophilic

Glands

   Irregular, variable size, some dilated

   Branching, infolding and outpouching

   Simple hyperplasia

      Haphazardly spaced in abundant stroma

   Complex hyperplasia

      Closely spaced with decreased stroma

      Highly irregular outlines

Frequent associated features

   Polypoid growth

   Ciliated cells

   Ectatic venules

Breakdown and bleeding

- Sometimes the glands are somewhat crowded and irregular but not densely packed, and it is not clear whether the process should be termed simple or complex hyperplasia. When the distinction between complex and simple hyperplasia is not clear, classify the lesions as simple hyperplasia.

Morphologic features of endometrial hyperplasia with atypia

Cytologic featuresa

    Nuclei

        Stratification with loss of polarity

        Enlarged, rounded with irregular shapes

        Coarsening of chromatin creating a vesicular appearance

        Prominent nucleoli

        Mitotic activity, variable amount

    Cytoplasm

        Eosinophilia, diffuse or focal

    Glands

        Irregular, variable size, some dilated

        Simple atypical hyperplasia

            Haphazardly spaced in abundant stroma

        Complex atypical hyperplasia

            Closely spaced with decreased stroma

            Highly irregular outlines

    Frequent associated features

        Papillary infoldings into glands (no bridging)

    Decreased stroma

    Ciliated cells

    Squamous metaplasia

- A diagnosis of atypical hyperplasia is best limited to those cases in which clearly atypical nuclei are readily identified without diligent searching. Atypia not be diagnosed unless clearly atypical nuclei involve most of the epithelium, lining several well-visualized glands in cross section. Surface epithelium should be avoided in assessing the presence of atypia. Atypia cannot be reproducibly subdivided or graded into categories such as mild, moderate, and severe.

Epithelial Cytoplasmic Change (Metaplasia)

- Epithelial cytoplasmic alterations, commonly designated metaplasia, often occur in the endometrium. The term "metaplasia" refers to transformation of cells to a type not normally found in an organ. By this definition, most of the alterations commonly classified as endometrial metaplasia do not qualify as such. Consequently, some of the cytologic transformations of the epithelium previously referred to as endometrial metaplasia are better classified as a "change." The latter term has the advantage of offering a descriptive designation without implying a specific mechanism of development. Because these "changes" are especially common in hyperplasia, it is important that they be recognized and clearly separated from more significant glandular abnormalities.

- There are five general types of cytoplasmic transformation that occur in the endometrium. These are squamous, ciliated cell, eosinophilic, mucinous, and secretory (clear cell and hobnail cell) change. [55]

Squamous differentiation:

- The nonkeratinizing morules have a characteristic appearance, forming solid nests of bland eosinophilic cells that fill gland lumens. The cells have uniform, round to oval nuclei with small nucleoli and rare or absent mitoses. The nuclei are centrally placed in dense, eosinophilic cytoplasm. When the squamous change forms morules, the gland is largely filled with a round to oval mass of uniform cells with indistinct cell borders. The intraglandular nests of squamous epithelium may show central necrosis, but this feature has no effect on the diagnosis or prognosis of the lesion.

Ciliated cell change:

- It is not a true metaplasia, as ciliated cells are normally present along the surface epithelium, being most numerous in proliferative endometrium. These cells often are interspersed in small groups among nonciliated columnar cells, but sometimes they are extensive and line most of the gland. Ciliated cells have pale to eosinophilic cytoplasm. The luminal border of these cells may show a cuticle of dense cytoplasm formed by the ciliary basal bodies. The nuclei are mildly stratified, yet they remain cytologically bland with round to oval shapes, an even chromatin distribution, and small nucleoli. The rounding and slight nuclear enlargement that characteristically occurs should not be considered as evidence of atypia. Mitoses generally do not occur in ciliated cells.

- **Eosinophilic (pink) cell change:**

  - This change actually represents several types of cytoplasmic transformation. Eosinophilic cells may be a variant of ciliated cells, squamous cells, or oncocytes as well as eosinophilic syncytial change.

  - Eosinophilic cell change that resembles ciliated cell change is common. In this situation the cells are columnar or slightly rounded and have a moderate amount of pale pink cytoplasm, resembling the cytoplasm of ciliated cells but lacking luminal cilia.

  - Eosinophilic cell change also merges with squamous change in some cases; in these instances the cells are rounded to polygonal and pavement-like, resembling cells seen in squamous differentiation but lacking the solid, morule-like growth pattern. In other cases eosinophilic cells contain abundant, granular cytoplasm resembling oncocytes or Hurthle cells seen in other organs.

  - Eosinophilic cell change may even show interspersed cells with a small amount of cytoplasmic mucin, suggesting overlap with mucinous cell change. In all these forms of eosinophilic cell change, the nuclei are often round rather than oval and somewhat stratified. Luminal cell borders are sharply demarcated. The nuclei are smaller and more uniform and lack the irregular nuclear membrane, chromatin condensation along the membrane, and prominent nucleoli that characterize cells with true cytologic atypia.

- **Mucinous change:**

  - It is characterized by the presence of abundant mucinous cytoplasm, resembling normal endocervical glandular cells.

  - The epithelium is also thrown into small papillary projections. This pattern is not as common as the other cytoplasmic changes and is seen most often in association with atypical hyperplasia or carcinoma. These cells are columnar, with basal nuclei and abundant pale supranuclear cytoplasm that contains mucin. Histochemical stains, such as mucicarmine or periodic–acid Schiff with diastase digestion, demonstrates the abundant cytoplasmic mucin.

- **Secretory and clear cell change:**

  - It is very infrequent once progestin-related effects are excluded. This is usually a focal alteration, limited to scattered glands. As the names imply, the cells contain clear, glycogen-rich cytoplasm and resemble those found in secretory or gestational endometrium. Rarely the cells develop a hobnail pattern with nuclei that protrude into the gland lumen, resembling the Arias-Stella reaction. The secretory/ clear cell change usually occurs in endometrium that shows estrogenic effects that range from a proliferative pattern to carcinoma.

  - Diffuse secretory changes sometimes occur in hyperplasia, and this has been called "secretory hyperplasia". This process can be seen in the premenopausal or perimenopausal patient with hyperplasia who has sporadic ovulation or who has been treated with progestins.

ENDOMETRIAL INTRAEPITHELIAL NEOPLASIA (EIN)

- EIN is the histopathologic presentation of a monoclonal endometrial premalignant glandular lesion prone to malignant transformation into endometrial (type I) endometrial adenocarcinoma. The pre cancerous properties of EIN are ones previously ascribed to atypical endometrial hyperplasia, a common ground which is the basis for managing EIN lesions in a manner similar to that of atypical endometrial hyperplasia. EIN is detected almost always within the context of an endometrial biopsy performed in response to patient symptoms or when monitoring women receiving HRT. Postmenopausal bleeding or vaginal bleeding or irregular menses in the perimenopausal period are the most common signs. The postmenopausal patient with a non cyclic atrophic endometrium may experience symptomatic bleeding directly from an EIN lesion. The average of women with EIN is 52 years which is about 8 years earlier than the average age of 60 for endometroid endometrial adenocarcinoma in the same patient population.

Cancer outcomes in women with EIN:

- Untreated EIN, whether diagnosed objectively by histomorphometry or subjectively using the table given below has a high likelihood of progression to adenocarcinoma. EIN lesions that do not progress to cancer may involute, or stably persist for protracted periods of times. Of import, 39% of patients with EIN had cancer diagnosed within the first year.

ETIOLOGY AND NATURAL HISTORY:
EIN is a monoclonal neoplasm:

- EIN lesions begin as localized monoclonal outgrowths of mutated endometrial cells. The clonal nature of EIN lesions has been demonstrated by various markers such as non-random X chromosome inactivation and clonal propagation of altered microsatellites.

Role of estrogens and progestins:

- Hormonal risk factors for EIN are the same as those previously described for atypical endometrial hyperplasia, with estrogens acting as promoters and progestins as protectors. Hormonal and genetic mechanisms are linked in the very earliest stages of endometrial carcinogenesis, through the selective effects of hormones upon genetically defective compared to intact endometrial cells. Hormones act upon 'latent' precancers, which are somatically mutated, histologically unremarkable, endometrial glands detectable only with specialized biomarkers.

MORPHOLOGY OF EIN:

- Most EIN lesions are themselves grossly inapparent. They expand by interactively remodelling the stroma relative to the neoplastic glands. This makes the boundaries difficult to distinguish grossly. One circumstance in which EIN may be grossly evident is when a thin atrophic background endometrium lacks the bulk necessary to contain the expanding EIN lesions without distortion. For this reason, some EIN lesions in postmenopausal patients are visible as local thickenings. Many EIN lesions are focally distributed against a background of benign endometrial hyperplasia, which can have a thickened and multi cystic appearance. EIN may present within otherwise grossly unremarkably sessile or pedunculated endometrial polyps. Thoroughness of sampling is a key element in successful detection of EIN lesions especially those that are physically small and localized at the time of diagnosis.

- EIN diagnostic criteria: This includes shared features conserved amongst all lesions and exclusion of benign mimics and carcinomas. All five EIN diagnostic criteria must be met in every case to maintain a high level of diagnostic specificity and clinical predictive value.

Diagnostic features of EIN

Architecture	• Area of glands exceeds that of stroma (glands/ stroma >1).
	• Lesion composed of individual glands which may branch slightly and vary in shape.
Cytology	• Nuclear and/or cytoplasmic features of epithelial cells differ between architecturally abnormal glands and normal background glands.
	• May include change in nuclear polarity, nuclear pleomorphism or altered cytoplasmic differentiation state.
	• If no normal glands present, highly abnormal cytology.
Size	• Maximum linear dimension exceeds 1mm
Exclude mimics	• Benign conditions with overlapping criteria
	Disordered proliferative
	Basalis
	Secretory
	Polyps
	Repair
Exclude cancers	• Carcinoma if maze like glands, solid areas, or significant cribriforming.

- Subjective (non-morphometric) EIN diagnosis from H&E stained slides at a standard microscope without specialized equipment, using the above table also has a high level of clinical outcome predictive valve.

CARCINOMA OF ENDOMETRIUM

- Carcinoma of the endometrium is the most common gynaecologic malignancy in developed countries. It typically occurs in elderly individuals 80% of the patients being postmenopausal at the time of diagnosis. Given uterine bleeding the probability of carcinoma is a strong function of the patient's age. The rate is 9% for women in their 50's; 16% for those in their 60s; 28% for those in their 70s and 60% for those in their 80s.

- Patients with endometrial adenocarcinoma fall into two loose clinicopathologic clusters. Patients in the first group (type I) tend to be between 40 and 60 years of age. These patients may have a history of chronic anovulation or estrogen hormone replacement therapy and the carcinomas are usually well differentiated Stage I, non myoinvasive tumors associated with endometrial hyperplasia. Most of the tumors are ER and PR positive and p53 negative and express low levels of the proliferation antigen Ki-67. Patients in the first group have a very favourable prognosis after hysterectomy.

- In contrast, patients in the second group (type II) tend to be elderly and typically have no history of hyperestrogenism. In these cases the surrounding non neoplastic endometrium is almost always atrophic and there is often an in situ component with high grade cytologic features. The carcinomas that develop in this group of patients are usually of the special variant type with a poor prognosis or high grade endometroid neoplasms that are first found with deep myoinvasion. The tumors tend to be ER/PR negative, strongly express p53, and show high Ki-67 labelling. The most common type of endometrial carcinomas endometroid adenocarcinoma; may be manifested by such clinical findings as obesity, infertility and late menopause. In particular high estrone and albumin bound estradiol levels were associated with increased risk in postmenopausal women than in premenopausal women. Diabetes has been repeatedly associated with an increased risk of endometrial cancer ranging from 1.2 to 2.1. [57]

CLASSIFICATION OF ENDOMETRIAL CARCINOMAS

The current International Society of Gynecological Pathologists and World Health Organization classification of endometrial carcinoma is mentioned below.

I) Endometrial adenocarcinoma Variants
    A. With squamous differentiation
        I. Adenocarcinoma with squamous metaplasia (adenocanthoma)
        Ii. Adenosquamous carcinoma
    B. Villoglandular
    C. Secretory
    D. Ciliated cells
II) Serous adenocarcinoma
III) Clear cell adenocarcinoma
IV) Mucinous adenocarcinoma
V) Squamous cell carcinoma
VI) Undifferentiated carcinoma
VII) Others
VIII) Mixed

ENDOMETROID ADENOCARCINOMA:

- Endometrioid adenocarcinoma is defined as a cancer in which the glandular pattern, when well differentiated has cytologic features most like a normal proliferative endometrium. The most common sub type of endometrial cancer is endometroid, its pure form constituting about 60% of all endometrial carcinoma. Most neoplasms develop slowly in the setting of hyperestrogenism against a background of benign endometrial hyperplasia and EIN, although some arise in atrophic endometrium. It is predominantly a disease of sixth and seventh decades and 75% of cases occur after the menopause. Until recent years, the diagnosis has been made by endometrial biopsy or curettage, but imaging techniques now play a part. Outpatient endometrial sampling techniques generally have a good diagnostic rate for endometrial carcinoma.

Morphology:

- Endometrioid carcinoma can present variously, the uterus may be slightly or grossly enlarged but it may be of normal size or even small and atrophic, particularly in a postmenopausal women. Most of them arise in the corpus, presenting either as single mass or two or three separate masses or a diffuse thickening of the endometrium. More frequently carcinomas are situated on the posterior wall. The most common appearance is of a raised, rough, perhaps papillary area of the endometrium with a shaggy surface and ulceration, frequently occupying at least half of the surface area of the endometrium. Sometimes the tumor is polypoid with a fairly narrow base. Myometrial invasion may be obvious to the naked eye, with either pushing or infiltrating borders but frequently it is difficult to appreciate the degree of myometrial invasion grossly. A tumor diameter of more than 2 cm generally is associated with a poorer prognosis and a higher frequency of distant failure.

- Microscopically, the glandular pattern and cellular features generally resemble that of the proliferative phase endometrium. Multilayered epithelial cells are nearly always seen. Occasionally cribriform fragments have a microglandular appearance easily confused with a cervical lesion. Solid growth may vary widely in extent a feature of importance in tumor grading. [57] The individual epithelial cells are larger than would be expected in the proliferative phase. The carcinoma cells have a distinctly altered cytology that varies between patients and even within areas of a single tumor, but may include rounded nuclei, clumped chromatin, and prominent nucleoli. Individual tumors frequently demonstrate patchy changes in differentiation to mucinous, squamous, tubal or other cytologies and in these cases cytoplasmic as well as nuclear features stand out from the normal background. Some endometroid adenocarcinomas secrete abundant extracellular mucin but lack much intracytoplasmic mucin a feature that distinguishes them from mucinous adenocarcinoma. Mitotic figures are usually present but may be scant in well differentiated tumors. The endometrial stroma adjacent to the newly forming well differentiated neoplastic glands responds by remodelling rarely demonstrating a classic desmoplastic change. For this reason, qualitatively assessing the character of the endometrial stroma is non contributory in resolving premalignant from malignant disease within the endometrial compartment itself.

Carcinoma is recognized within the endometrial compartment by presence of at least one of the following:
- A cribriform pattern of the glands.
- A solid mass of glandular epithelium.
- Meandering interconnected lumens formed by folded sheets of neoplastic epithelium.
- Irregular, angulated and tapering glandular contours.

- Foamy histiocytes are commonly seen in the endometrial stromal compartment of patients with a carcinoma. Nearly a fifth of cases contain stromal cells laden with lipid but there is no correlation between the presence of these cells and the grade of the tumor or the survival of the patient. This change is simply a reactive response to tumor cells that have died.

Phase of endometrioid endometrial carcinogenesis:

- A dynamic model of endometrial tumor regimens based on mutation driven sequential clonal selection during tumor evolution resembles those proposed for several other tumor systems. Each sequential clonal selection occurs through successful competition with elements of the parent field. As such, clonal proliferation in tissues like the endometrium may be viewed as evidence of the creation of a cell or group of cells with a growth advantage – a characteristic of neoplastic processes.

- Presumptive identification of the earliest stages of endometrioid endometrial (type I) precancer has now been accomplished using a variety of molecular marker systems derived from this model. These include lineage continuity (forward carry over) of mutations of genes such as PTEN and Kras, between premalignant and malignant carcinogenesis phases in individual patients, and demonstration of monoclonal growth of premalignant tissues. [58]

MODIFIED FIGO SURGICAL STAGING AND GRADING SYSTEM FOR UTERINE CORPUS CARCINOMA

Stage I: Confined to the uterine corpus
- IA : Tumor limited to endometrium
- IB : Invasion of less than half of the myometrium
- IC : Invasion of more than half of the myometrium.

Stage II: Uterine cervix involved
- IIA : Endocervical glandular involvement only
- IIB : Cervcial stroaml invasion

Stage III: Pelvic extension
- IIIA : Tumor invasion of serosa and/or adnexa and/or positive peritoneal cytology.
- IIIB : Vaginal metastasis
- IIIC : Metastases to pelvic and/or para-aortic lymph nodes.

Stage IV: Extrapelvic extension
- IVA : Tumor invasion of bladder and/or bowel mucosa.
- IVB : Distant metastases including intra-abdominal and/or inguinal lymph nodes.

HISTOLOGIC GRADING SYSTEM:
Degree of architectural differentiation

Grade 1: 5% or less of a non squamous or non morular solid growth pattern.

Grade 2: 6% - 50% of a nonsquamous or non morular solid growth pattern.

Grade 3: >50% of a non squamous or non morular solid growth pattern.

NUCLEAR GRADING:
The nuclear grade is determined by the variation in nuclear size and shape, chromatin distribution, and size of the nucleoli.

Grade 1 nuclei are oval mildly enlarged and have evenly dispersed chromatin

Grade 3 nuclei are markedly enlarged and pleomorphic with irregular, coarse chromatin and prominent esoinophilic nucleoli.

Grade 2 nuclei have features intermediate to grades 1 and 3.

- Mitotic activity is an independent histologic variable, but it is generally increased with increasing nuclear grade, as are abnormal mitotic figures. [59]

OTHER VARIANTS OF ENDOMETIROID ADENOCARCINOMA:

Four histologic variants of endometrioid adenocarcinoma are recognized which are –

i. Endometrioid adenocarcinoma with squamous differentiation.
ii. Endometrioid adenocarcinoma with secretory differentiation.
iii. Endometrioid adenocarcinoma with ciliated cell differentiation.
iv. Endometrioid adenocarcinoma with villoglandular variant.

Endometrioid adenocarcinoma with squamous differentiation:

- One fourth of endometrial adenocarcinomas display focal squamous differentiation. In the late 1960s the distinction was made between tumors where the squamous component was 'well' or 'poorly', differentiated. The former tumors, were called 'adenocanthoma' and the latter 'adenosquamous' carcinoma. Numerous studies have confirmed the significantly better prognosis associated with adenocanthoma and endometrial carcinoma without squamous differentiation (about 90% survival at 5 years) as compared to adenosquamous carcinoma (65%), a distinction related largely to the grade of the glandular component. Usable criteria are needed for the recognition of squamous differentiation so that a solid focus of adenocarcinoma will not be mistaken for squamous differentiation.

- The International Society of Gynecological Pathologists (ISGP) has suggested criteria in response to this need.

Criteria for identifying squamous differentiation in endometrioid adenocarcinoma

Keratin or Keratin pearls demonstrated without special stains

Inter cellular bridges

Atleast three of the following

    Sheet like growth without gland formation or pallisading

    Distinct cell margins

    Deeply eosinophilic or 'glassy' cytoplasm

    Decreased nuclear: cytoplasmic ratio compared with the rest of the tumor.

- In large studies where the glandular and squamous components were graded independently, differentiation of the squamous component was shown to parallel that of the glandular component closely. Thus, the prognosis can be predicted by the glandular grade alone. Grading of the glandular component is also superior in predicting lymph node metastasis and 5 year survival. [57]

Endometrial adenocarcinoma, endometrioid type, with secretory differentiation:

- 'Secretory adenocarcinoma' is an uncommon variant of endometrioid adenocarcinoma composed of well differentiated glands resembling the early or mid-secretory endometrium. The most common changes are subnuclear and/or supranuclear vacuolation. In addition, there may be solid areas consisting of small polygonal cells with clear cytoplasm. Secretory adenocarcinomas are associated with a good prognosis.

- The distinction of secretory carcinoma from atypical hyperplasia with secretory effect can be difficult and is based on the presence of stromal invasion in the carcinoma. [59]

Endometrial adenocarcinoma, endometrioid type with ciliated cell differentiation:

- Ciliated cell carcinoma is a rare form of differentiation in low grade endometrioid carcinoma. It does not need to be classified separately from endometrioid carcinoma. Patients range in age from 42 to 79 years, are often postmenopausal and present with bleeding. Microscopically, ciliated carcinoma is almost always well differentiated and often displays a cribriform pattern. The gland lumens in the cribriform areas are lined by cells with prominent eosinophilic cytoplasm and cilia.

Endometrial adenocarcinoma, endometrioid type, villoglandular variant:

- Villoglandular carcinoma is a variant of endometrioid carcinoma that displays a papillary architecture in which the papillary fronds are composed of delicate fibrovascular core covered by columnar cells that generally contain bland nuclei. The median age is 61 years, similar to that of women with typical endometrioid carcinoma. The microscopic appearance of villoglandular carcinoma is characterized by thin, delicate fronds covered by stratified columnar epithelial cells with oval nuclei that generally display mild to moderate (grade 1 or 2) atypia. Mitotic activity is variable. Myometrial invasion is usually superficial. The main consideration in the differential diagnosis is serous carcinoma because both villoglandular and serous carcinoma have a prominent papillary pattern.

MUCINOUS ADENOCARCINOMA:

- An endometrial carcinoma composed predominantly of cells containing prominent intracytoplasmic mucin, resembling the mucinous tumors found in the endocervix. This comprises 1-9% of all endometrial adenocarcinoma. To qualify as a mucinous carcinoma, more than one half the cell population of the tumor must contain periodic acid Schiff (PAS) positive, diastase – resistant intracytoplasmic mucin.

- This type of endometrial carcinoma is seen in patients aged between 47 and 89 years and typically present with vaginal bleeding.

- No macroscopic features distinguish mucinous from endometrioid adenocarcinoma, apart from the infrequent prominence of the secreted tenacious mucus. The most frequent architectural pattern is glandular, often in a villoglandular configuration. The involved cells are tall with basal nuclei and prominent intracytoplasmic mucin.

- Mucicarmine, PAS and Alcian blue are all useful to amplify the staining. Cribriform areas are unusual, cystically dilated glands filled with mucin and papillary fronds surrounded by extracellular lakes of mucin containing neutrophils are typical. Curiously mucinous differentiation is sometimes associated with squamous differentiation. Nuclear atypia is mild to moderate and mitotic activity is not prominent.

- The distinction of mucinous carcinoma of the endometrium from clear cell or secretory carcinoma is made on the basis of morphology and PAS and mucin stains. The cells in secretory carcinoma are clear (not granular or foamy) because of the presence of glycogen which is PAS positive and is removed by diastase treatment. Rarely, mucinous carcinoma may contain areas that simulate microglandular hyperplasia of the cervix. Such foci are characterized by cells showing mucinous and eosinophilic change with microcystic spaces containing acute inflammatory cells. The patients with microglandular hyperplasia are younger. When stratified by stage, grade, and depth of myometrial invasion, mucinous tumors behave as do endometrioid carcinomas. However they are of low grade and minimally invasive and therefore as a group have an excellent prognosis. [57]

CLEAR CELL CARCINOMA

- An endometrial adenocarcinoma exhibiting clear or sometimes eosinophilic cells and 'hobnail' shaped cells arranged in a papillary, tubulocystic or solid pattern, similar to the clear cell adenocarcinoma that occur in the vagina, cervix and ovary. This type accounts for 1-6% of endometrial carcinomas and occur at an older age (mean 65-69 years). The women generally are less often obese, less often have diabetes mellitus.

- **Morphology:** There are no gross features that distinguish clear cell adenocarcinoma from other varieties of endometrial carcinoma but. Clear cell adenocarcinoma which occur in the vagina, cervix and ovary as well as endometrium have similar histologic appearances and are striking and unmistakable. In order of decreasing frequency the most common patterns are papillary, glandular, solid and tubulocystic. Most clear cell adenocarcinomas have a mixture of atleast two of these patterns. The epithelial cells lining the cysts in the tubulocystic areas frequently contain very little cytoplasm, so that the enlarged and pleomorphic nuclei appear to protrude into the lumen, presenting a 'hobnail appearance'. However the most prominent diagnostic feature of the clear cell adenocarcinoma is clear cytoplasm in many cells. Nothing in the cytoplasm stains with H&E since the glycogen that is present is leached out during normal fixation and processing. The stroma may be dense and hyalinized, particularly in the tubulocystic areas. Nuclear pleomorphisim is often marked and the tumors are of a high nuclear grade. The frequency of mitotic figures is variable. Architectural grading is difficult, nuclear grade may be used, but the recent WHO classification considers clear cell adenocarcinomas as high grade by definition.

- Clear cell adenocarcinoma has to be separated particularly from secretory and serous endometrial carcinoma as well as from yolk sac tumor. Clear cell and serous carcinoma may at times have areas that are indistinguishable. Serous carcinoma is rarely solid and does not show a Tubulocystic pattern. There is evidence that p53 immunostaining is less pronounced in clear cell compared to serous carcinoma.

- Clear cell adenocarcinomas may at times be exceedingly difficult to distinguish from Arias Stella change in the endometrium. Arias Stella change is often associated with deciduas and in women who are usually young. The nuclear chromatin is virtually always smudged in Arias Stella change whereas the chromatin material in atleast some areas in clear cell adenocarcinoma should be crisp.

- The prognosis for women with endometrial clear cell adenocarcinoma is poorer than for endometrioid carcinoma with 5 and 10 year disease free survival rates reported as 43-68% and 39% respectively. Pathologic stage and age are the two most important prognostic factors. [57]

SEROUS ADENOCARCINOMA

- Serous adenocarcinoma is an aggressive form of endometrial cancer exhibiting a predominantly papillary architecture composed of exfoliative bullous hobnail like cells with marked nuclear atypia. The term 'papillary serous carcinoma' is more commonly used synonymously in the literature. Serous adenocarcinoma is a poor prognostic form of endometrial cancer. It accounts for 1-10% of all endometrial cancers.

- Serous carcinomas generally have the same gross features as endometrioid carcinomas, although many appear as bulky, necrotic masses. The uterus is more frequently atrophic as compared with endometrioid carcinomas. Histologically, they typically have complex, branching papillae with usually broad, thick fibrovascular cores, but occasionally thin to delicate cores. The papillae are covered by a stratified epithelium with a prominent and very characteristic tufting or budding pattern with many groups of detached cells lying free between the papillae. They may also show glandular or solid pattern, where the glandular structures are irregularly shaped and often lined by polygonal cells. The nuclei are usually high grade, with marked pleomorphism and large macronucleoli along with occasional bizarre and hyperchromatic giant nuclei. Psammoma bodies are found in about 25% of these cases. Majority of cases exhibit striking lymphovascular invasion and deep myometrial invasion and a higher incidence of cervical and lower uterine segment involvement. Serous adenocarcinoma of endometrium is considered high grade by definition.

SQUAMOUS CELL CARCINOMA

- An endometrial carcinoma composed entirely of malignant squamous epithelium, similar to squamous cell carcinoma found elsewhere, squamous cell carcinoma of the endometrium is rare. The median age of patients is 61 years. Many cases are associated with benign squamous metaplasia of the endometrium (ichthyosis uteri), a condition encountered only in older women, often in association with intrauterine infection.

There are no gross features that distinguish squamous cell carcinoma from endometrioid adenocarcinoma. The histologic features resemble that of squamous cell carcinoma found in other sites. It has long been believed that the diagnosis should be made only if there is no evidence of coexistent adenocarcinoma and if careful examination of the cervix excludes a primary tumour. Where squamous cell carcinoma is present synchronously in both the uterine body and cervix, the assumption has been that the tumor originated in the cervix and spread upwards. More recent views suggest that sometimes the opposite may be true. The prognosis is exceedingly poor, with 26% of reported cases surviving only a median of 9 months after diagnosis. [57]

MESENCHYMAL TUMORS

ADENOFIBROMA:

- Adenofibroma is a benign neoplasm that usually arises in the cervix or endometrium of mainly postmenopausal women. A tumor of mixed mullerian origin, this neoplasm is papillary and most often sessile and the stromal component predominates. This tumor is exceedingly rare and must be differentiated from the more common adenosarcoma with a subtle malignant stromal component. Grossly the tumor occupies and distends the uterine cavity as a broad based polypoid mass often lobulated. A bland epithelium that is usually of endometroid type but which may be endocervical, ciliated or even squamous covers broad or fine papillary stromal fronds. Cells of fibroblastic type and more rarely endometrial stroma or smooth muscle make up the mesenchymal element. The stroma may be cellular or fibrous. Mitotic figures are usually absent and should not exceed 1 per 10 hpfs in the most active areas. The stromal elements are cytologically bland without nuclear pleomorphism. [61]

CARCINOSARCOMA:

- Uterine carcinosarcoma, previously referred to as malignant mixed mullerian tumors are aggressive neoplasms composed of epithelial elements and mesenchyme both of which are histologically malignant. The sarcomatous component may be composed of homologous or heterologous tissue. They are often diagnosed at an advanced stage which contributes significantly to the poor 5 years overall survival rate of less than 35%. Carcinosarcoma accounts for 2-5% of all malignancies of the uterine corpus. It is most common in postmenopausal women (median age 66 years) but has also been reported in younger women and children. The most common presentation is abnormal vaginal bleeding or abdominal pain or an abdominal mass may also be the first symptoms. Carcinosarcoma shares risk factors with endometrial carcinoma, but the influence of these factors is weaker. Some carcinosarcomas have also been known to possess estrogen and progesterone receptors. [61]

- Carcinosarcomas lack a characteristic naked eye appearance. Some arise in atrophic uteri while others develop into enormous masses infiltrating and expanding the uterine wall. More typically, it forms a broad based polyp that fills and expands the uterine cavity often protruding through the cervix. The cut surface may show extensive necrosis and hemorrhage, and gritty or hard areas may be present corresponding to bone or cartilage. Myometrial invasion may be obvious grossly in most cases.

- Microscopically, both the epithelial and mesenchymal elements are malignant, the carcinomatous component corresponding to any mullerian type. Most frequent epithelial component is a poorly differentiated serous carcinoma. Endometrioid differentiation is less common. Others like squamous cell carcinoma, mucinous carcinoma or clear cell carcinomas may rarely be encountered, sometimes the sarcomatous component may predominate. The stromal elements may be homolgous or heterologous but usually is obviously malignant. The homologous components commonly found are endometrial stromal sarcoma, fibrosarcoma and undifferentiated sarcoma. The heterologous elements most commonly observed are rhabdomyosarcoma, chondrosarcoma, osteosarcoma and liposarcoma. Carcinosarcomas typically contain cells with striking variations in size and shape, bizarre mitotic figures and giant cells. There is often extensive necrosis, deep myometrial invasion and cervical involvement with lymphovascular permeation and often extra uterine spread.

HISTOGENESIS OF CARCINOSARCOMAS:

- There are several possible ways in which carcinosarcomas may originate. The 'conversion' theory is currently widely accepted, where the sarcomatous elements are thought to arise from the carcinoma during the tumors evolution Tumors arising by this mechanism should be monoclonal in contrast to collision tumors. Evaluation of p53 and K-ras mutations as well as immunohistochemical and cytogenetic studies has confirmed that most carcinosarcomas are monoclonal. [62]

- Disease stage has been the most consistent independent predictor of outcome in patients with uterine carcinosarcoma. Other clinicopathologic features associated with worse outcome in early stage CS include deep myometrial invasion; lymphovascular space involvement, histology of the carcinomatous component, extent of the sarcomatous component, and the presence of heterologous element. There is a growing body of clinical and pathologic data suggesting that uterine carcinosarcomas are more similar to high-risk endometrial adenocarcinomas than to high grade uterine sarcomas.

- These inlude risk factors for development of disease; predilection for lymphatic and intraperitoneal dissemination; presentation at advanced stage and response to similar chemotherapy.

- The most important entities in the differential diagnosis of carcinosarcoma are mullerian adenosarcoma with stromal over growth, endometrioid adenocarcinoma with prominent spindle cells, carcinomas with undifferentiated component and monophasic pleomorphic tumors. [63]

- **Altrabulsi B. Et al.**, studied clinicopathologic features of 16 cases of endometrial undifferentiated carcinoma and compared them with 33 cases of endometrioid adenocarcinoma FIGO-G3. The differential diagnosis of undifferentiated carcinoma should include in addition to high grade endometrioid adenocarcinoma, high grade sarcomas and malignant mixed mullerian tumor. The distinction between endometrioid type grade 3 and undifferentiated adenocarcinoma is important owing to the worse prognosis of the latter. [64]

- **Alkushi A. Et al.,** proposed a novel system for grading of endometrial carcinoma and comparison with existing grading system. The presence of atleast 2 of three criteria: predominantly papillary or solid growth pattern, mitotic index ≥ 6/10 HPF or severe nuclear atypia would result in a tumor being considered high grade. Low grade tumors satisfy at most one of these criteria. [65]

- **Nofech-Mozes S. Et al.,** reviewed 827 case of pure endometrial endometrioid adenocarcinoma and concluded that endometrioid adenocarcinoma manifests most commonly with low tumor grade and without deep myometrial invasion. High tumor grade is significantly associated with deep myhometrial invasion, cervical involvement and LVI. [66]

- **Soslow RA et al.** studied 187 high grade endometrial cancers consisting of Endometrioid (EC-3) carcinoma – grade 3 FIGO, Serous carcinoma (SC) Clear cell carcinoma (CC). High grade endometrial cancers of different histologic sub types treated in an individualized manner areassociated with similar clinical outcomes, but differences in age at presentation, race, distribution, association with hyperplasia, stage and sites of tumor dissemination support the ideas that these represent distinct disease entities as defined by traditional histopathologic classification of endometrial cancers. [67]

AUB in case of Pregnancy and Abortion [68]

- Recognition of the features of gestational endometrium, trophoblast, and villi, as well as the pathologic changes in chorionic tissues, is an important part of endometrial biopsy interpretation. The presence of intrauterine products of conception generally excludes the diagnosis of ectopic pregnancy and can help explain other pathologic states such as abnormal bleeding or chronic endometritis.

- An abortion before the 16th week of pregnancy is the usual source of endometrial tissue specimens that show apparent gestational changes. The different types of abortions are defined as follows. "Spontaneous abortions" are unexpected and unplanned interruptions of pregnancy that present with bleeding and passage of tissue. Approximately 15% to 20% of early pregnancies end in a spontaneous abortion.

- Besides abortions, several other complications of pregnancy, such as retained placenta or placental implantation site, ectopic pregnancy, or gestational trophoblastic disease, lead to the need for endometrial curettage.

Endometrial Glands and Stroma in Pregnancy:
Early Gestational Endomatrium (1 to 3 weeks post fertilization)

- These changes include recrudescence or accentuation of glandular secretions, distension of the glands, edema, and an extensive predecidual reaction. The coiled glands show secretory activity and a serrated lumen, but they appear distended or wider than those in the late secretory phase of a menstrual cycle. Vascular prominence with engorgement and dilation of superficial veins and capillaries also occurs, and the spiral arteries develop thicker walls.

- Within 10 to 15 days of fertilization, the endometrium gradually begins to show more characteristic changes of pregnancy as differentiation of stromal cells into decidua progresses. As compared to predecidual cells, decidual cells are larger and contain more abundant eosinophilic to amphophilic cytoplasm that may contain faint vacuoles. These cells become more clearly polyhedral with well-defined cell membranes. Nuclei of the decidualized stromal cells are round to oval and uniform, with smooth outlines, finely dispersed chromatin, and indistinct nucleoli. Occasional decidualized stromal cells are binucleate. Stromal granular lymphocytes persist in early pregnancy and are clearly evident among decidual cells. The presence of granular lymphocytes may suggest a chronic inflammatory infiltrate, but the granular lymphocytes, in contrast to inflammatory cells, have characteristic lobated nuclei and plasma cells are not present.

Endometrial Glands and Stroma in later Pregnancy:
(4 or more weeks post fertilization)

- With advancing gestational age, pregnancyrelated patterns become more pronounced and distinctive. The decidualized stromal cells are widespread and prominent, especially as the cell borders become better defined, and they develop an epithelioid appearance.

- The decidua shows small foci of physiologic necrosis during pregnancy, as it remodels during growth of the fetus and placenta and as the decidua capsularis fuses with the decidua parietalis. The hypersecretory pattern of the glands begins to regress early in pregnancy, and with increasing decidualization the glands become atrophic. Conversely, in areas where the glands appear hypersecretory, the stroma often is not decidualized. Usually a mixture of hypersecretory and atrophic glands is present. By the end of the first trimester, the glands for the most part are atrophic and have lost their luminal secretions. In fact, as they form irregular, dilated spaces with indistinct epithelium, they may be difficult to distinguish from vascular channels. As pregnancy advances, the spiral arteries maintain thick walls, a feature that persists to term and is helpful in recognizing gestational changes. In the first trimester the arteries develop a characteristic atherosclerosis like change when an intrauterine pregnancy is present, characterized by subintimal proliferation of myofibroblasts with foam cells.

Arias–Stella Reaction:

- At 4 to 8 weeks after blastocyst implantation, the endometrium often shows at least a focal Arias–Stella reaction in the glands. This glandular change is a physiologic response to the presence of chorionic tissue either in the uterus or at an ectopic site. The morphologic features of the Arias–Stella reaction include nuclear enlargement up to three times of normal size and nuclear hyperchromasia, often accompanied by abundant vacuolated cytoplasm. The cells typically are stratified and the nuclei hobnail-shaped, bulging into the gland lumen. These large nuclei may contain prominent cytoplasmic invaginations. Mitotic figures are rarely present.

- This process may be extensive, involving many glands or the reaction can be focal, and involving only a few glands. The changes can even be limited to part of a gland, leaving the remaining nuclei unaffected. The Arias–Stella reaction has two histologic patterns. One is a "hypersecretory" change characterized by highly convoluted glands lined by cells with stratified nuclei and abundant clear to foamy cytoplasm. The other pattern has been termed "regenerative," although this hypothesized etiology for the change remains unsubstantiated. This pattern is characterized by glands lined by enlarged hobnail cells with little cytoplasmic secretory activity. In fact, the two patterns are not very distinct and there is frequent overlap between them.

- This change occurs as early as 4 days after implantation, although it generally is seen after about 14 days. The Arias–Stella reaction persists up to at least 8 weeks following delivery.

Other Glandular changes in Pregnancy:

- Besides the Arias-Stella reaction, the endometrial gland cells may undergo other specific changes in the presence of trophoblastic tissue. One such change is abundant clear cytoplasm; gland cells accumulate abundant amounts of clear, glycogen-rich cytoplasm. The nuclei in areas of clear cell change can become stratified, which, combined with the abundant clear cytoplasm, can result in apparent obliteration of the gland lumens.

- Another pregnancy-related change is optically clear nuclei of gland cells; caused by accumulation of a filamentous material in the nuclei.

- In early pregnancy, endometrial glands become strongly immunoreactive for S-100 protein. This immunoreactivity rapidly disappears after the 12th week of gestation.

Trophoblast and Villi

- In early pregnancy trophoblastic proliferation begins with the development of the blastocyst, the outer layer of which is termed the trophoblastic shell. Villous formation does not begin until about 7 days after implantation of the blastocyst (13 days following conception). For morphologic identification, the products of conception are divided into three components: (1) the villi and their trophoblast ("villous" trophoblast), (2) the implantation site ("extravillous" trophoblast), and (3) fetal tissues. Usually these tissues are easy to recognize.

Trophoblastic cells

- The trophoblast is extraembryonic but fetal in origin, growing in intimate association with host maternal tissues. Very early in pregnancy trophoblastic cells differentiate and invade decidua, even before villi form.4 At this stage of early gestation, implanting trophoblast is the predominant component of placental tissue. The trophoblast continues to grow along this interface of maternal and placental tissue throughout pregnancy. The decidua basalis where trophoblast interfaces with the endometrium and myometrium becomes the placental implantation site. The trophoblastic cells are the epithelial component of the placenta and are divided into three cytologically and functionally distinct populations: cytotrophoblastic (CT) cells, syncytiotrophoblastic (ST) cells, and intermediate trophoblastic (IT) cells.

- CT cells are the germinative cells from which other trophoblastic cells differentiate. Accordingly, they are mitotically active. They are uniform cells about the size of a decidualized stroma cell, with a single nucleus, one or two nucleoli, pale to faintly granular cytoplasm, and prominent cell borders.

- ST cells are larger and multinucleate with dense amphophilic to basophilic cytoplasm. The nuclei of ST cells are dark and often appear pyknotic; they do not contain mitoses. The cytoplasm also typically contains small vacuoles and larger lacunae in which maternal erythrocytes can be identified. A microvillous brush border sometimes lines the lacunae of the ST cells. CT and ST cells typically display a dimorphic growth pattern, with the two cell types growing in close proximity.

- The intermediate trophoblast develops from cytotrophoblast on the villous surface, and in early pregnancy is manifested as sprouts and columns that extend to and extensively infiltrate the underlying decidua at the implantation site. In fact, the predominant location of the IT is at the implantation site, which explains why it is often called "extravillous cytotrophoblast."

- Another older term for IT is "X cells."[4] The IT actually represent a heterogeneous population of trophoblastic cells: the villous IT, the implantation site IT, and the chorionic-type IT. The morphologic and immunohistochemical features of these IT subtypes depend on their differentiation status and their anatomic location.

- The IT that extends from the trophoblastic column of the anchoring villi is designated "villous" IT. These cells are mononucleate and larger than CT cells. They have pale cytoplasm and large, round nuclei.

- The implantation site IT cells infiltrates the decidua and the myometrium and have heterogeneous appearance.

- The chorionic-type IT constitute the cells of the chorion laeve where they form a cohesive layer of epithelium, are composed of relatively unform cells with eosinophilic to clear (glycogen-rich) cytoplasm. These cells are smaller than implantation site IT although an occasional cell is multinucleated.

Chorionic Villi and Villious Trophoblast in First Trimester

- Villus formation in very early pregnancy depends on the existence of the embryonic disc. Villi begin to develop on the 12th to 13th day postfertilization, and by days 12 to 15 the placenta can develop for a while independently, without the presence of an embryo. In the early stages of pregnancy, villi have a loose, edematous stroma with few welldeveloped capillaries. Once the yolk sac and embryo develop, vascular circulation is established in the villous stroma and these vessels contain nucleated red blood cells.

- A few histologic changes in the placenta help to determine the age of the developing conceptus.

Events in first trimester placental development

Blastocyst implantation	6–7 days
Villus formation beginsb	12 days
Nucleated rbcs from yolk sac appear in villi	4.5 weeks
Non-nucleated rbcs from liver appear in villi	5–6 weeks
Proportion of nucleated rbcs decreases from 100% to 10% in villi	4.5–9 weeks
Decrease in prominence of inner cytotrophoblast layer	16–18 weeks

Hydropic Change and Other Pathologic Changes in Abortions:

- The microscopic features of the decidua and the products of conception in curettage samples vary depending on the type of abortion. Villi are usually normal in therapeutic abortions, whereas they tend to reflect early death of the embryo in spontaneous or missed abortions. Therapeutic abortion specimens may show pathologic changes in the villi, however. In spontaneous or missed abortions, placental morphology is influenced by gestational age, karyotype, and regressive changes. With the death of the embryo, the villi often show hydropic change because of loss of the villous vascular supply, especially if embryonic death occurs very early, often before 4.5 weeks postfertilization age. The avascular villi are mildly distended with fluid and the curettage samples do not contain fetal tissue giving the changes of the so-called blighted ovum. This Pattern of mild villous edema and no evidence of fetal development indicate that the embryo either never developed or ceased development at a very early stage of gestation. Microscopically, villous edema in a hydropic abortion can appear especially prominent. Hydropic change affects most villi but is minimal and microscopic; cistern formation in the villi is rare.

- Other morphologic changes in chorionic villi from first-trimester abortion specimens may be found. Irregular outlines of villi yielding a scalloped appearance and trophoblastic invagination into the villous stroma forming pseudoinclusions often are associated with abnormal karyotypes of the conceptus.

- In missed abortions the villi often are necrotic or hyalinized and the deciduas are necrotic. Another change in villi associated with death of the embryo is loss of villous vascularity and fibrosis of the villous stroma.

GESTATIONAL TROPHOBLASTIC DISEASE

- Gestational trophoblastic disease (GTD) includes disorders of placental development (hydatidiform mole) and neoplasms of the trophoblast (choriocarcinoma, placental site trophoblastic tumor [PSTT], and epithelioid trophoblastic tumor). The recent classification of these lesions by the World Health Organization (WHO) clearly defines the different histologic forms of GTD.

Modified World Health Organization classification of gestational trophoblastic disease

Molar lesions
    Hydatidiform mole
        Complete
        Partial
    Invasive mole
Nonmolar lesions
    Choriocarcinoma
    Placental site trophoblastic tumor
    Epithelioid trophoblastic tumor
    Miscellaneous trophoblastic lesions
    Exaggerated placental site
    Placental site nodule
Unclassified trophoblastic lesion

Hydatidiform Mole

- Hydatidiform mole, either partial or complete, is infrequent in the United States and Europe, occurring in about one in 1000 to one in 2000 pregnancies, although some studies have suggested that partial mole may be even more frequent, occurring in up to 1 in 700 pregnancies. In other parts of the world, including Asia and Latin America, these disorders are more common, although problems in methodology often complicate studies of their frequency when deliveries take place at home.

- The separation of Hydatidiform Mole into two subtypes: complete and partial. Both forms have different cytogenetic and clinicopthologic profile.

Invasive Hydatidiform Mole

- In invasive hydatidiform mole, hydropic molar villi and hyperplastic trophoblast either invade myometrium or are present at other sites, usually the vulva, vagina, or lungs. To establish the diagnosis, it is necessary to clearly identify molar villi beyond the endometrium. In curettings this requires finding the villi within myometrial smooth muscle, an extremely rare event. Consequently, invasive mole is almost never diagnosed by endometrial biopsy or curettage. It is important to remember that the presence of residual mole in a recurettage specimen does not represent invasive mole in the absence of demonstrable myometrial invasion.

Trophoblastic Neoplasms

Choriocarcinoma

Gestational choriocarcinoma can occur in the uterine cavity following any type of pregnancy. As a rule, the risk of choriocarcinoma increases with the abnormality of the antecedent gestation. Complete hydatidiform mole is a major predisposing factor, and about half the cases of choriocarcinoma follow a complete mole. Choriocarcinoma also can arise from the trophoblast of an abortion or a term pregnancy. Consequently, this lesion may be present whenever abnormal vaginal bleeding occurs during the postpartum period in a young woman who has had a pregnancy of any type. Patients with choriocarcinoma also can present with metastatic disease without uterine signs or symptoms. Typically, the patient with choriocarcinoma has markedly elevated serum hcg titers.

- Choriocarcinoma is hemorrhagic and necrotic, composed of trophoblastic cells without villi that invade normal tissues.

- It has two main diagnostic criteria, the fisrt criterion, and absence of villi, is important, as the proliferative trophoblast of hydatidiform moles or of early normal pregnancy can closely simulate the trophoblast of choriocarcinoma.

- The second criterion of choriocarcinoma, a dimorphic pattern of syncytiotrophoblastic (ST) cells alternating with nests or sheets of mononucleate trophoblast (cytotrophoblastic [CT] or intermediate trophoblastic [IT]) cells should be found, at least focally, to establish a histologic diagnosis of choriocarcinoma.

- The admixture of ST with CT or IT cells yields a plexiform pattern. In these cases identification of ST cells is an important diagnostic feature. These cells contain multiple nuclei, ranging from 3 to more than 20 per cell, which are variable in size. Often the nuclei are pyknotic but they can be vesicular with prominent nucleoli. ST cells have dense eosinophilic to amphophilic cytoplasm with small vacuoles or large lacunae that often contain erythrocytes. CT cells are small (about the size of a decidualized stromal cell) and uniform. They have a single nucleus with a prominent nucleolus, pale to clear cytoplasm, and distinct cell borders. Large IT cells with polygonal shapes and one or two large, hyperchromatic nuclei.

Placental Site Trophoblastic Tumor

- The placental site trophoblastic tumor (PSTT) is a rare form of trophoblastic neoplasia composed predominantly of implantation site IT. The hcg titer is generally low and may not be noticeably elevated if a sensitive assay method is not used. Because pstts extensively infiltrate the myometrium, the uterus can be perforated during curettage. These neoplasms usually are benign, despite destructive growth in the myometrium.

- PSTT typically produces a mass lesion. These tumors range from focal lesions 1 to 2cm in diameter to large masses that replace much of the corpus. Curetting of PSTT typically yields multiple fragments of neoplastic tissue.

- Microscopically PSTT is composed predominantly of implantation site IT cells that invade normal tissues. These cells generally are polyhedral and grow in cohesive masses that often show areas of necrosis. The IT cell cytoplasm is generally amphophillic with occasional clear vacuoles and indistinct cell borders; nucleus is pleomorphic, hyperchromatic; nucleoli have deep folds or grooves and others may have pseudoinclusions. Binucleated and multinucleated IT cells also present.

- PSTT is diffusely immunoreactive for human placental lactogen (hpl) and Mel-CAM (CD 146). Ki-67 labeling index may be a significant prognostic indicator as it is usually greater than 50% in malignant tumors but only about 14% in benign PSTT.

Epithelioid Trophoblastic Tumor

- Epithelioid trophoblastic tumor is a rare form of trophoblastic tumor. This trophoblastic neoplasm is distinct from choriocarcinoma and PSTT with features resembling those of somatic carcinomas. The epithelioid trophoblastic tumor is preceded by a term gestation in two thirds of cases, with spontaneous abortions and hydatidiform moles. Usually there is a long interval following the gestation and the diagnosis of this tumor with a range of 1 to 18 years. Serum hcg levels are usually elevated at the time of diagnosis, although the levels are generally low (<2500miu/ml).

- Epithelioid trophoblastic tumor is composed of chorionic-type IT. ST cells are indistinct.The tumor displays a nodular growth pattern and has a striking epithelioid appearance, both in its cytologic features and in its pattern of invasion.The neoplasm is composed of small nests and cords of cells. The nests often contain dense central hyaline material and necrotic debris, and the cords are encompassed by a hyaline matrix.The predominant cells are relatively uniform in size and are mononucleate with round, uniform nuclei and eosinophilic or clear cytoplasm. They are larger than CT cells but smaller than implantation site IT cells. Apoptotic cells and islands of necrotic debris are abundant in most tumors. This tumor is diffusely reactive for cytokeratin (AE1/AE3 and cytokeratin 18) as well as epithelial membrane antigen.

Placental Site Nodule

- Placental site nodules are small, circumscribed foci of hyalinized implantation site with IT cells that occasionally present in an endometrial biopsy or curettage. The lesion itself is circumscribed, nodular, or plaque-like with densely eosinophilic, hyalinized stroma containing aggregates of IT cells. Often focal chronic inflammation including plasma cells surrounds the nodule, while the rest of the endometrium shows no inflammation. The trophoblastic cells in these nodules resemble chorionic-type IT. In the placental site nodule the cells vary in size; many have small, uniform nuclei and some larger cells show irregular, hyperchromatic nuclei. Occasional multinucleated cells are present. Mitoses are rare or absent.

Exaggerated Placental Implantation Site

- The exaggerated placental site is characterized by focal finding maintaining architecture, an increase in the number and size of individual IT cells. In addition, widely dispersed multinucleated IT cells are a component of the trophoblastic infiltrate. Often several fragments of tissue in curettage samples contain portions of the lesion, and this process can extensively infiltrate fragments of myometrium. A few chorionic villi may be present. IT cells appear larger and more hyperchromatic than normal. Despite their apparent prominence, these IT cells show no mitotic activity.

MYOMETRIAL PATHOLOGY

INTRODUCTION

- The bulk of the myometrium is composed of smooth muscle cells. Uterine smooth muscle tumours may be divided into benign-leiomyomas, malignant leiomyosarcomas and those tumours of uncertain malignant potential. [69] Smooth muscle tumours of uterus show a wide spectrum of histology and biological behaviour. Patterns of growth, histological appearance, associations with vessels and degenerative changes provide the basis for the classification of most benign smooth muscle tumours of uterus. [70] Diagnostic problems are encountered when these lesions exhibit unusual histological features that mimic leiomyosarcomas or endometrial stromal tumours. [71]

- Leiomyomas are the commonest tumours found in women. They occur in 20 – 25 % of women over the age of 30 years. [72] Leiomyosarcomas are the most frequent pure sarcomas to arise in the uterus, which account in most studies for between 25 and 45 % of uterine sarcomas and upto 1% of all uterine malignancies. [73] The present study is proposed to be undertaken, because myometrial tumours continue to be a major cause of morbidity and leading indication for hysterectomy in premenopausal women. And some of the leiomyomas such as mitotically active leiomyomas, bizarre leiomyomas may simulate leiomyosarcoma, as these tumours carry, better prognosis, requires detailed morphological study.

EMBRYOLOGY

- Uterus arises from the paramesonephric ducts. They appear in the embryo at about 40 days. In the female embryo, as soon as the paramesonephric ducts come into deposition within the urorectal septum and begin to fuse, the uterus is being formed.

- Around the 19th week the corpus begins to differentiate into layers of mucosa, muscle and serosa. The muscular wall of the uterus condenses from the outer layer of the fused terminal ends of paramesonephric ducts to encase the endometirum and its canal cavity. A well marked fundus is apparent at 26 weeks and the change in the form of the upper limit of uterus from a V-shaped notch to a convex curve is due to the general thickening of its walls brought about by the growth and development of muscle tissue. [73, 74, 75]

ANATOMY

- The myometrium is a fibromuscular layer that forms most of the uterine wall. In the nulliparous women, it is dense and 1.3 cm thick at the uterine mid level and fundus but thin at the tubal orifices. It is composed largely of smooth muscle fascicule mingled with loose connective tissue, blood vessels, lymphatics and nerves. [74]

- The body of the uterus is often described as having four or less distinct muscular layers. The inner most layers (submucosal layer) is composed mostly of longitudinal and some oblique smooth muscle. Where the lumen of the uterine tube passes through the inner wall, this layer forms a circular muscle coat. External to the submucosal layer is the vascular layer, a zone rich in blood vessels as well as longitudinal muscle. Next is a layer of predominantly circular muscle, the supravascular layer. The outer thin longitudinal muscle layer, the subserosal layer, lies adjacent to the perimetrium. [74]

HISTOLOGY

- The myometrium consists of compact bundles of smooth muscle separated by thin strands of interstitial connective tissue, blood vessels, lymphatics and nerves. The bundles of smooth muscle are seen in cross, oblique, and longitudinal sections. The smooth muscle cells are of uniform size, contain moderate eosinophilic fibrillar cytoplasm. The nuclei are elongated, cigar shaped, having finely dispersed chromatin. [71, 73, 74]

CLASSIFICATION

- The vast majority of smooth muscle neoplasms of the uterus are benign. There are, however a number of variants of smooth muscle tumours that by virtue of their gross features, cytological appearances or their unusual growth patterns, cause diagnostic difficulties, as they overlap with malignant neoplasms and cause major difficulty in reliably predicting clinical out come. [73] No single morphological feature is reliable in predicting the biological behaviour. Therefore a multivariate approach is required. [73] The classification system, which takes into account mitotic activity, cellular atypia, the presence or absence of coagulative tumour cell necrosis and the histological type of neoplasm, is the best presently available. [76]

A comprehensive classification of smooth muscle tumours of the uterus developed by the **World Health Organization (WHO) is as follows:**

Smooth muscle tumors
Leiomyosarcoma
Epitheloid variant
Myxoid variant
Smooth muscle tumour of uncertain malignant potential
Leiomyoma, not otherwise specified.
Histological variants
Mitotically active variant
Cellular variant
Haemorrhagic cellular variant
Epitheloid
Myxoid
Atypical variant
Lipoleiomyoma variant
Growth pattern variants
Diffuse leiomyomatosis
Dissecting leiomyoma
Intravenous leiomyomatosis

Metastasizing leiomyoma
Miscellaneous tumours
Adenomatoid mesothelioma
Other tumors [77]

LEIOMYOMA

Epidemiology:

- Leiomyomas are the most common tumours found in women. They occur in 20 – 25 % of women over the age of 30 years. [71, 72, 73] They are uncommon in women younger than 30 years, although occasional cases have been recorded in teenagers. [73]

- Leiomyomas are more common in black than white females with a roughly threefold difference being confirmed. The vast majority of leiomyomas occur in uterine corpus.

- In Gynaecological practice in England, it is generally stated that an eighth to a tenth of all patients suffer from leiomyoma. An incidence of similar range was given in America and Egypt. [78]

Histiogenesis:

- Histological examination of seedling fibroid suggests that they develop from smooth muscle cells of the myometrium. [78] After ultrastructural study of minute uterine leiomyomas, it has been proposed that the progenitor cells of leiomyomas may be undifferentiated mesenchymal cells in the uterus that proliferate and differentiate into smooth muscle cells under certain pathologic conditions. [79]

Aetiology:

- The cause of uterine leiomyoma is undetermined. Many factors have been considered on theoretical grounds but practical support is lacking. Heredity is considered to be a factor "fibroids run in families". From the clinical and epidemiological surveys in the USA, it is known that fibroids are 3-9 times more common in Negroes than in Caucasians. It has been suggested that, this is due to the higher prevalence of pelvic infections in black women, with abnormal uterine growth occurring secondarily to the myometrial irritation caused by these infections. An alternative view may be the presence of a gene encoding for fibroid development, as there is often a positive family history of fibroids in patients who develop these tumours. [72]

- They are true neoplasms rather than hyperplastic proliferations. This has been convincingly demonstrated by analysis of iso-enzymes of glucose-6 phosphatedehydrogenase. [73]

- The factors involved in the initiation and growth of leiomyomas remain poorly understood. They probably involve complex interactions of sex steroid hormone and local growth factors with somatic mutations in the normal myometrium. Uterine leiomyomas appear during the reproductive years and regress after the menopause, indicating ovarian steroid-dependent growth potential. Further evidence for the role of female sex hormones in the growth of leiomyomas is their occasional rapid growth and haemorrhagic degeneration associated with pregnancy, clomiphene and progestagen treatment. [72, 73] Oestrogen and progesterone promote the development of myofilaments and dense bodies' ultrastructural features of smooth muscle differentiation. [73]

Steroid Growth Factors:

- A number of studies have investigated oestrogen and progesterone as possible growth factors in fibroids. Two types of nuclear oestrogen (E2) binding sites have been found in myometrium and fibroid muscle. Type I nuclear receptors bind oestrogen with high affinity and low capacity, and type II receptors bind oestrogen with low affinity and high capacity. [72]

- There is significant elevation of oestrogen and progesterone receptors content in fibroids compared to normal myometrium. [72, 79, 80] Though oestrogens seem to support active growth of these tumours, growth does not always coincide with increased oestrogens and they may sometimes continue after climacteric. [81]

Peptide Growth Factors:

- Epidermal Growth Factor (EGF) is a single chain polypeptide of 53 amino acids, has a mitogenic property, and mediates the role of oestrogen in fibroid growth. TGF-α, insulin-like growth factors, fibroblast growth factor (FGF) are other growth factors which may act either alone or in combination. IGF I and II correspond to somatomedin C and A respectively. They are produced in response to circulating growth hormone, may have a local growth stimulating effect on fibroids. FGF has fibroblast as one of its targets. As fibroids are partly composed of fibrous tissue, FGF may be important for their growth. [72]

CYTOGENETICS:

- Uterine leiomyomas are benign neoplasms that exhibit clonal chromosomal abnormalities. Approximately 40% of uterine leiomyomas; ave chromosomal abnormalities detectable by conventional cytogenetic analysis, including t (12:14) (q15:q23-24), rearrangement involving the short arm of chromosome 6, interstitial deletion of the long arm of chromosome 7, rearrangement of 10q22, rearrangement of 13q 21-22, deletions of 3q. [73, 82]

IMMUNOHISTOCHEMISTRY:

- Smooth muscle cells, in the myometrium and within smooth muscle tumours react with antibodies to muscle specific actin, alpha smooth muscle actin, desmin and caldesmon. There is immunoreactivity with vimentin, but the intensity of staining and the proportion of cells that stains are less than the muscle specific antibodies.

- Cytokeratin immunoreactivity is frequently observed in myometrium and in the smooth muscle tumours, the extent and intensity of reactivity depends on the antibodies used and the fixation of the specimen. Epithelial membrane antigen is usually negative. [73, 82]

CLINICAL FEATURES:

- The clinical features of uterine fibroids are variable. The vast majority are symptomless especially when small. The symptomatology and severity usually depend on the size, position and number of fibroids present. However most leiomyomas are believed to be asymptomatic and progress slowly. They may be multiple and may remain asymptomatic regardless of location. [72, 83]

Menstrual Abnormalities:

- Approximately 30% of women with fibroids have been reported to have menstrual abnormalities, most often menorrhagia. [72, 73, 83] Menorrhagia may occur when cavity surface area is expanded by submucous fibroids. However often submucous fibroids are not present, but extensive uterine bleeding exists. The increased bleeding maybe due to either increased vascularity of the uterus or anovulatory cycles.

- Fibroids arising at various sites in the uterus could cause congestion and dilatation of endometrial venous plexuses by impinging and obstructing veins in the myometrium.

- The resultant obstruction could cause endometrial venule ectasia which may play a role in enhanced uterine bleeding. Intermenstrual bleeding may occur when a fibroid polyp is undergoing necrosis, and when a pedunculated fibroid is extruded through the cervical canal. [72]

Pain and Pressure Effects:

- Chronic dull backache maybe present when the fibroid is of moderate size in a retroverted uterus. Dysmenorrhoea may also occur. Acute pain may be present with red degeneration, necrosis and torsion of a pedunculated fibroid and with usage of the oral contraceptive pills. Pressure symptoms depend on the area of impingement of the fibroid and its size – there may be bladder, bowel or vascular symptoms. [72]

Fibroids and Pregnancy:

- The role of fibroids as a causal factor in infertility remains controversial. It is obvious that obstruction of both uterine tubes by fibroids or gross uterine cavity distortion could contribute to infertility. However, fibroids are common and certainly occur in both normally fertile and infertile women and there is no clear evidence that the mere presence of fibroid is causally linked to infertility especially when small and not impinging on the uterine cavity. The size of the fibroid is the most important prognostic factor. [72]

- There have been a number of theories as to the cause of infertility in women with fibroids. Distortion of the endometrial cavity, greater distance for sperm travel, impairment of blood supply to the endometirum, atrophy and ulceration of the endometrium. [72] There have been associations with preterm delivery, abnormal presentation, outlet obstruction, postpartum haemorrhage and puerperal sepsis. [72, 82]

Parasitic Leiomyoma:
- Subserosal leiomyomas when pedunculated may undergo torsion, hemorrhage, infarction and detachment from the uterine serosa, resulting in a parasitic leiomyoma, deriving its blood supply from new vascular anastomoses outside the uterus. Infection of leiomyoma may occur leading to fever and leucocytosis. [72, 73]

GROSS APPEARANCE:

- The usual leiomyoma occurs within the myometrium as a well circumscribed, solid white tumour, which maybe single or multiple. They bulge above the surrounding myometrium from which they are easily shelled out. Cut surface are white to tan with a whorled trabecular pattern and may show various patterns of degeneration. [71, 72, 73, 82]

MICROSCOPIC FEATURES:

- Leiomyomas are typically composed of smooth muscle cells with bland, uniform, cigar shaped nuclei arranged in interlacing bundles, showing little or no mitotic activity. Nuclear chromatin is finely dispersed. The smooth muscle cells are separated by viable amounts of collagen, which increases in amount with age. Most leiomyomas are more cellular than surrounding myometrium. The borders are usually sharply defined from surrounding myometirum. [71, 73, 76, 78, 82] Increased mitotic activity is associated with secretory phase of the menstrual cycle. [79]

FIBROIDS IN DIFFERENT SITES:

- Leiomyoma originates in the myometrium and remains there as an intramural fibroid in half of cases. It may draw inwards to bulge into the cavity of the uterus as a submucous fibroid or grow outwards to distort the surface of the uterus as a subserosal fibroid. Rarely a tumour grows between the folds of the broad ligaments – intraligamentary fibroid. Fibroids usually present in more than one of these situations in the same specimen. [78] Leiomyomas are more common in fundus. These regional differences in myometrial susceptibility to the development of leiomyomas may be due to differential responsiveness of hormones or growth factors or secondary to precursor lesions like myometrial hyperplasia, metaplasia or dysplasia. [80]

TYPES OF DEGENERATIVE CHANGES:

- Degenerations are frequent in leiomyomas and usually are due to inadequate blood supply. Degeneration, though often central first, may be diffuse. Fibroids at any site are subject of degenerative change but those with a pedicle are particularly susceptible. The nature of the change varies according to the degree and rapidity of onset of vascular insufficiency and is found in one or several of a group of fibroids. In some instances degeneration is associated with acute and severe symptoms. [78]

- All these features maybe encountered in malignant tumours and therefore thorough sampling of lesions with unusual gross appearance is imperative. Further, in the presence of extensive degeneration, it is difficult to identify the smooth muscle nature of the lesion. [71, 73, 78, 82]

a) Atrophy:
- Gradual shrinkage of a fibroid usually follows cessation of ovarian function. The muscle fibers of the tumor become harder in consistency and may calcify. [78]

b) Hyaline Degeneration:
- Commonest type of degeneration develops slowly and results from the tumour outgrowing its blood supply. The tumours become softer in consistence, white or yellow in colour. In large tumours the hyaline tissue liquefies producing cystic areas. Deposition of collagen initially within stroma, separating smooth muscle bundles and later replacing the smooth muscle cells with few or no surviving nuclei. [71, 73, 78]

c) Cystic Change:
- Cystic change is merely a sequel to hyaline degeneration. [78] Occurs most frequently in subserosal leiomyomas, is less common in intramural tumours, and is least common in submucosal tumours. Fluid accumulation may result in extremely large tumours. The cyst fluid is usually straw coloured, red or brown discoloration of the cyst content is usually a reflection of intracystic haemorrhage. [84] Initially irregular strands of tissue line the cavities on histological examination; the wall of the cystic area is composed of hyalinised leiomyoma.

d) Myxoid degeneration:
- Grossly appears as gelatinous or watery cut surface with pools of mucoid material, usually limited and focal, may be generalized. Histologically, myxoid change refers to the presence of scattered nuclei embedded in an amorphous, pale staining slightly bluish fibrillar matrix. [71, 73, 78, 84]

e) Haemorrhagic Cellular Degeneration:
- Manifested as central red softening. Histological examination shows cellular, mildly pleomorphic and mitotically active zones around the areas of haemorrhage or hyalinization / infarction. [71, 73, 78]

f) Red Degeneration:
- Refers to tumour with gross appearance of red beef pattern with haemorrhage. Histological examination shows extensive infarction sometimes with surviving marginal zone of benign smooth muscle cells. [71, 73, 78]

g) Calcification:
- Common in fibroids of postmenopausal women, and also is liable to occur following necrosis. Calcification occurs in two forms. In one form there are granular gritty masses throughout the tumour. These coalesce and spread into the capsule, the contour of the tumour becoming irregular. [78]

- In the other, calcification is peripheral and a brittle many layered shell is produced. This second type is a sequel to diffuse necrosis, particularly red degeneration. Histologically appears as purplish amorphous lake with haematoxylin. Ossification may appear in a calcified area. [78]

h) Hydropic Degeneration:
- Hydropic degeneration refers to the accumulation of abundant edema fluid, typically associated with variable amounts of collagen. It is usually a focal finding on both gross and microscopic examination of otherwise typical uterine leiomyomas. [82] Rarely can be seen following torsion. The fluid is derived from liquefaction of hyaline material. [83]

i) Infection:
- Infection of a leiomyoma usually occurs as a sequel to torsion or in association with acute endometritis. [78]

j) Necrosis:

- Grossly appears as sharply demarcated yellow areas. There are three types of necrosis. Hyaline necrosis; coagulative tumour cell necrosis and ulcerative necrosis in submucosal lesions. The presence or absence and type of necrosis are powerful predictors of clinical behaviour.

- In coagulative tumour cell necrosis, there is an abrupt transition between necrotic and preserved cells. The hematoxyphilia of the nuclei is often retained in the necrotic cells and there is no associated inflammation. The characteristic low power microscopic pattern is one of the blood vessels cuffed by viable cells surrounded by a sea of necrotic tumour. Coagulative tumour cell necrosis is commonly present in clinically malignant smooth muscle neoplasms. In contrast, hyalinising necrosis has a distinctly zonal pattern with central necrosis, a more peripheral zone of granulation tissue, and at the periphery, a viable amount of hyaline eosinophilic collagen interposed between the central degenerated region and peripheral preserved smooth muscle cells. Ulcerated submucous leiomyomas features acute inflammatory cells and an associated peripheral reparative process. [71, 72, 76, 82]

k) Telangiectasis:

- Cystic spaces containing blood may be seen in a fibroid. They are merely distended blood vessels of tumour possibly due to mechanical obstruction to blood flow. [78]

l) Lymphangiectasis:

- May occur due to distension of lymphatics. The cysts are lined by endothelium. [78]

LEIOMYOMA VARIANTS

HISTOLOGICAL VARIANTS:
Mitotically Active Leiomyoma:
- The term is used for leiomyomas exhibiting a slight increase in cellularity, with insignificant cytological atypia, an absence of coagulative necrosis and increased mitotic count of 5 to 19 per 10 high power fields. [71, 73, 82, 85, 86] The age of the patient with 'mitotically active leiomyoma' or 'leiomyoma with increased mitotic index' is relatively young (mean age – 39 years). [86] They can vary in size from 0.6 cm to 20 cm and more frequently found in a submucosal location, where they may be subjected to traumatic stimuli. [73] Increased mitotic activity is associated with secretory phase of the menstrual cycle, pregnancy, the use of exogenous hormones. Microscopic evidence of haemorrhage or necrosis maybe seen. [87]

Cellular Leiomyoma:
- Cellular leiomyoma is one of the most common variant of leiomyoma, which has been defined by the World Health Organization as a leiomyoma that is significantly more cellular than adjacent myometrium, with the notation that this subtype correctly diagnosed should account for ≤5% of leiomyomas by the pathologist. [88] These leiomyomas characteristically have low mitotic indices almost always less than five mitotic figures per 10 High Power Field (HPF). Those very occasional cellular leiomyomas with the mitotic indices in excess of five mitotic figures per 10 HPF are best termed "mitotically active cellular leiomyomas". [73]

- "Highly cellular leiomyoma' have been recently defined as a distinct entity, the cellularity of which is as great as that of typical endometrial stromal tumors. [73, 77, 82, 88] These are commonly confused with either stromal nodules or low grade endometrial stromal sarcomas.

- The patient ranged in age from 29 to 65 years (Mean 46). The most common clinical presentation was irregular uterine bleeding, usually menorrhagia. [88] Grossly, the lesions are generally intramural, solid, circumscribed.

Microscopically highly cellular leiomyomas demonstrate large blood vessels with thick muscular wall, cleft like spaces, merging of cells with the myometrium peripherally, absence of foam cells. [88]

- In contrast, endometrial stromal neoplasms are characterized by a diffuse growth of small oval cells and a network of small calibre, thin walled arterioles with the addition of an infiltrating margin and vessel invasion in low grade stromal sarcomas. Immunoreactivity for desmin is helpful in differentiating cellular and highly cellular leiomyomas from stromal tumours. Diffuse desmin immunoreactivity supports smooth muscle differentiation and focal immunoreactivity, or an absence of desmin reactivity supports endometrial stromal differentiation. [73, 82, 88]

Haemorrhagic Cellular Leiomyoma (Apoplectic Leiomyoma):
- Haemorrhagic cellular leiomyomas is a distinctive smooth muscle tumour occurring in women taking oral contraceptives or who are pregnant or recently postpartum. It was first described by Briscoe, illustrated by Hillard and Norris and termed "apoplectic leiomyomas" by Myles and Hart. [73, 89]

- Clinically the uterus may be sufficiently painful from rapid growth or tumour rupture. Some may present with haemoperitoneum due to rupture of the leiomyoma into the peritoneal cavity. It is not known if the association is related to the dose or duration of the oral contraceptive. Norethindrone, a progestogen known to have induced a haemorrhagic cellular leiomyomas when used alone. [73]

- Grossly, the uteri are enlarged and may contain single or multiple nodules with haemorrhage evident within them, sometimes associated with cystic change.

- Histologically – usually circumscribed, densely cellular and composed of oval to spindle cells with central haemorrhage and oedema. Areas of cellularity are seen usually around the haemorrhagic areas. Upto 20 mitotic figures / 10 HPF have been seen. Leiomyoma may show thrombosis, fibrinoid change, medial thickening, fibrosis or myxoid change. [73, 89]

Bizarre Leiomyoma (Atypical, Symplastic, Pleomorphic Leiomyoma):
- According to the WHO classification a bizarre leiomyoma is defined as "a leiomyoma containing giant cells with pleomorphic nuclei and little or no mitotic activity". [73, 77, 90] The distribution of the bizarre cells is found to be multifocal in a third and diffuse in half of the cases. Most workers consider extensive multifocal pleomorphism to be acceptable but truly generalized pleomorphism to be a worrying feature, usually associated with leiomyosarcoma. [5] These tumours are generally small <5.5 cm in 83% of cases, generally encountered in women of reproductive age with a range of 25 to 51 years. [71, 73, 90]

- Microscopically – contain bizarrely shaped multinucleated or mononucleated giant cells with hyperchromatic nuclei intermingled with cells showing usual smooth muscle differentiation. Intranuclear inclusions, cells with abundant, granular eosinophilic cytoplasm are seen. Coagulative tumour cell necrosis is not seen in any of these cases. [90] Other features noted in the study were the frequent presence of hyalinization, hypercellularity, fibronoid vascular change, chronic inflammatory cells, and fibrin thrombi. [90]

- Smooth muscle tumours with bizarre nuclei are all benign when there is no coagulative necrosis and widespread scattered foci of atypia, regardless of mitotic index, but leiomyosarcoma when atypia is diffuse and mitotic index in excess of 10 per 10 HPF. [76]

- Areas resembling bizarre leiomyomas may occur within or adjacent to leiomyosarcoma. Thorough sampling of uterine tumours is mandatory to exclude leiomyosarcoma. [90] Combination of aneuploidy and high MIB-1 activity strongly favours a diagnosis of leiomyosarcoma over that of leiomyomas with bizarre nuclei. [73]

Epitheloid Leiomyoma:

- This category includes leiomyoblastoma, clear cell leiomyomas and plexiform leiomyomas. [91] Epitheloid smooth muscle tumours have the same histologic appearance in the uterus as in other sites of the body. The mean age of women is in fifth decade with a range of 30-78 years. Grossly, yellow or gray in colour and may contain areas of haemorrhage. Majority are solitary, and they can occur in any part of the uterus. Median diameter of 6-7 cm.

- Microscopically, the cells are round to polygonal rather than spindle shaped, arranged in clusters or cords. The nuclei are round, relatively large and centrally placed. [82, 91, 92] There are three basic subtypes of epitheloid leiomyoma: leiomyblastoma, clear cell leiomyoma and plexiform leiomyoma. Mixture of the various patterns is common, providing the basis for designating all of these as epitheloid leiomyomas. Leiomyoblastoma is composed of round cells with eosinophilic cytoplasm rather than spindle cells.

- In clear cell leiomyoma, a subtype of epitheloid leiomyoma, cells are polygonal and have abundant clear cytoplasm with well defined cell membranes. The cells may contain glycogen but there is minimal lipid and mucin is absent. [82, 91]

- Plexiform leiomyoma is characterized by cords or nests of round cells with scanty to moderate amount of cytoplasm. Transition to more typical spindled smooth muscle cells is often identified within epitheloid leiomyomas.

- Small plexiform leiomyomas that are detected only on microscopic examination are referred to as plexiform tumour lets, usually solitary and submucosal, but they can occur anywhere in the myometrium and even in endometrium. [82]

- Epitheloid leiomyomas with circumscribed margins, extensive hyalinization and a predominance of clear cells generally are benign. The behaviour of epithelioid leiomyomas with two or more of the following features are not well established:

1) Large size (greater than 6 cm)
2) Moderate mitotic activity
3) Moderate to severe cytologic atypia
4) Necrosis.77

Myxoid Leiomyoma:
- Myxoid leiomyomas are benign smooth muscle tumours in which myxoid material separates the tumour cells. They are soft and translucent. Histologically, abundant amorphous myxoid material is present between the smooth muscle cells.

- The margins are circumscribed, and neither cytologic atypia nor mitotic figures are present. [73, 82]

- Leiomyoma with myxoid stroma is diagnosed, when cells are small and uniform, atypia is absent or atmost mild, Mitotic Index (MI) < 2 MF/10 HPF. Areas of ordinary leiomyoma should be present focally. Useful features in differentiating malignant from benign myxoid tumours are, tumour size, the nature of the margin, presence or absence of vascular invasion. [73]

"Perinodular Hydropic" Leiomyoma (Perinodular Hyaline Leiomyoma):
- Hydropic degeneration refers to the accumulation of abundant oedematous fluid, which is a common focal finding on gross and microscopic examination of otherwise typical leiomyomas. It is often associated with hyalinisation. [73]

- Hydropic degeneration can sometimes cause significant diagnostic confusion, in particular when it occurs in a perinodular distribution. These 'perinodular hydropic' or 'perinodular hyaline' leiomyomas characteristically possess well to poorly circumscribed zones of oedematous connective tissue. The hydropic areas typically consist of scattered fibroblasts set in a pale background with rounded clumps of collagen fibres. [73, 93] Where the hydropic / hyaline areas surround surviving nodules of smooth muscle tumour the perinodular pattern is produced and this may macroscopically give the tumour a "sago pudding" or 'in aspic' pattern. Numerous blood vessels, ranging from capillaries to large vessels with thick hyalinised walls, are a common finding in the hydropic and hyalinised areas. [71, 73, 93] The perinodular hydropic or hyaline pattern can, in some cases, be confused with intravascular leiomyomatosis or leiomyosarcoma (myxoid) because of hydropic change extending beyond the confines of leiomyoma.

Leiomyoma with Lymphoid Infiltration:
- Cases of lymphoid infiltration confined to Leiomyoma have been described. It may be focal to diffuse, moderate to markedly dense infiltrate of small lymphocytes. A scattering of plasma cells, eosinophils and occasionally prominent germinal centers may be seen. The main differential diagnosis is from a malignant lymphoma and inflammatory pseudotumour. [71, 73]

- Leiomyoma with lymphoid infiltration, on gross examination, resembles typical leiomyomas. Malignant lymphoma, in contrast, has a softer, fleshy, poorly circumscribed appearance. The lymphocytes in leiomyomas with lymphoid infiltration tend to be small and admixed with plasma cells and eosinophils, with the lymphoid infiltrate almost clearly confined to the leiomyomas. [71, 73]

- Inflammatory pseudotumour resembles grossly the leiomyomas, but there are significant histological differences. Inflammatory pseudotumours are composed predominantly of myofibroblasts and/or fibroblasts admixed with polymorphs. [71, 73]

Neurilemoma – like leiomyomas
- Uterine neoplasms composed of spindle cells, some of which exhibiting neurilemoma like palisading are designated neurilemoma like leiomyomas. Most such leiomyomas are benign: however, the prognosis is based on the

same mitotic and morphological criteria used for usual smooth muscle neoplasms. [85]

Lipoleiomyoma:
- The term lipoleiomyoma covers a spectrum of lesions with a fat content that ranges from being a minor component in what is otherwise a leiomyomas to being a neoplasm composed entirely of mature adipocytes. [85]

- They tend to occur in middle aged or elderly women, grossly, an otherwise typical leiomyoma is seen to have soft yellow areas within it. Histologically, the neoplasm comprises an admixture of varying amount of mature adipose tissue which can be seen as circumscribed areas or diffusely distributed within the leiomyomas. The origin of the adipose component appears to be by lipomatous metaplasia of a pre-existing leiomyoma. [71, 73, 82, 85]

- A morphologically closely related lesion is an angiolipoleiomyoma. It contains an admixture of mature fat, smooth muscle and blood vessels. The latter are seen as anomalous arterial blood vessels which show an irregular tortuous appearance resembling those seen in a renal angiomyolipoma. [73] Lipomatous component may be seen in other types of leiomyomas such as epitheloid leiomyomas and bizarre leiomyomas. [73] Pure lipoma of the uterus are very rare, only few cases exist in the literature. [73, 94]

Leiomyoma with Tubules:
- Tubular differentiation, in otherwise typical leiomyomas, has been described. The tubules were lined by a single layer of cuboidal cells. Diagnosis to be considered in a possible case of leiomyoma with tubular differentiation includes Mullerian adenosarcoma and carcinoma metastatic to a leiomyoma. [73]

- In Mullerian adenosarcoma the glands are benign and are seen throughout the tumor. They are frequently dilated and are surrounded by a rim of stromal condensation, so called "cambium layer". Adenosarcoma rarely contains smooth muscle as a major stromal component. Usually pleomorphic cells forming tumor glands in metastatic adenocarcinoma are easily recognizable as malignant and from a metastatic tumour. [73] "Adenomyoma" is also a differential diagnosis. They are characterized by endometrial glands within a localized mass of proliferating benign smooth muscle. [73, 77]

GROWTH PATTERN VARIANTS:

Dissecting Leiomyoma:

- Dissecting leiomyoma refers to a benign smooth muscle proliferation with a border marked by the dissection of compressive tongues of smooth muscle into the surrounding myometrium. [69, 70, 77, 95] They have been seen in women from third to early sixth decade of life. Three potential types have been described.

- The first type is the cotyledonoid dissecting leiomyomas of the uterus (the Sternberg tumor). This tumour is composed of exophytic masses of multinodular or multilobular tissue. Grossly, resembles placental tissue, often protruding from the lateral surface of the uterine cornu into the broad ligament and pelvic cavity. The exophytic components are in continuity with intramural dissecting tumour. It is the characteristic congested, cup shaped, red bulbous protuberances over the surface of the lesion that led the authors to use the name "cotyledonoid" (derived from Greek word meaning cup shaped hollow) and dissecting refers to the intramural component that irregularly extends into the myometrium between fascicles of normal smooth muscle cells. Histologically, the intrauterine component comprises fascicles of disorganized smooth muscle cells with a swirled appearance, marked vascularity and extensive hydropic degeneration. [70]

- The second type of dissecting leiomyomas is the so called "intramural dissecting leiomyomas", which is histologically similar but lacks the exphophytic component and is confined to the uterus. [96]

- More recently a third type of lesion has been described as "pure cotylednoid leiomyomas" i.e. an exophytic benign smooth muscle neoplasm extending into the peritoneal cavity and broad ligament but not associated with either a parent intramural mass or intramural dissection. This neoplasm apparently arose from the junction of the myometrium and serosa. [73]

<u>Intravascular (Intravenous) Leiomyomatosis:</u>

- Intravascular or intravenous leiomyomatosis has been reported in women ranging 27 to 80 years of age with most patients being middle aged females (median age – 44 years). The diagnosis is very rarely made prior to surgery. The intravenous component frequently involves the veins of the broad ligament but can also extend into the inferior vena cava and reach right side of heart. Coexisting leiomyomas of usual type are typically also seen within the myometrium. [73, 96]

- The main symptoms are abnormal bleeding and pelvic discomfort and pelvic mass. [82] When intravascular leiomyomatosis involves the right side of the heart, it produces symptoms of congestive cardiac failure or tricuspid valve obstruction. [73]

- Grossly, intravenous leiomyomatosis is a complex coiled or nodular growth within the myometrium with convoluted, worm like extensions into the uterine veins in the broad ligament or into the other veins. The growth extends into the venacava in more than 10% of patients, and in some cases it reaches as far as the heart. [71, 73, 82, 97]

- There are 2 main theories of origin of intravenous leiomyomatosis. According to one theory, the neoplasm arises from the wall of the veins; other theory is that it results from vascular invasion of the myometrium by a leiomyomas. [97]

- Histologically, appearance is variable, some examples are similar to leiomyomas, but most contain prominent zones of fibrosis or hyalinization.

Any variant smooth muscle histology, i.e. cellular, atypical or epitheloid may be encountered in intravenous leiomyoma. [73]

- The main differential diagnosis includes endometrial stromal sarcoma (low grade), leiomyomas with perinodular hydropic change and leiomyomas with vascular invasion. [71, 73, 97]

Diffuse Leiomyomatosis:
- Diffuse leiomyomatosis is a rare entity characterized by large number of coalescent, small benign smooth muscle tumours which diffusely involve and symmetrically enlarge the uterine cavity. [73, 77, 82, 85, 98, 99]

- Presentation is with abnormal uterine bleeding, lower abdominal or pelvic pain or an enlarged uterus. The age range in the literature is 22-38 years. Obstetrical problems may be associated with diffuse leiomyomatosis to a greater extent than the usual leiomyomas. [73, 98]

- The weight of the hysterectomy specimens range from 300 to 1260 grams. The hyperplasic smooth muscle nodules range from histological to three cm in size, but most are less than one cm in diameter. [9] Histologically, they are composed of uniform, bland, spindle shaped smooth muscle cells and are less circumscribed than leiomyomas. [77, 98] Diffuse adenomyosis bears a close macroscopic resemblance to diffuse leiomyomatosis, but presence of glands and stroma, on microscopy, easily distinguishes these two entities. [72, 73, 85, 98]

Diffuse Peritoneal Leiomyomatosis:
- Diffuse peritoneal leiomyomatosis is a rare condition characterized by the presence of multiple smooth muscle, myofibroblastic and fibroblastic nodules on the peritoneal surface of the pelvic and abdominal cavities of reproductive age. [73, 82, 85]

- Most cases are associated with pregnancy, an estrinizing granulosa tumour and oral contraceptive use. Most common presentation is as an unexpected finding at the time of cesarean section, lower abdominal pain or menorrhagia. [99]

- Grossly, appears as multiple small granular white to tan nodules on the pelvic and abdominal peritoneum, on the surface of uterus, adnexa, intestines and omentum. The nodules are distributed randomly and most of them are less than one cm in diameter. This contrasts with metastatic leiomyosarcoma, in which the nodules tend to be fewer, larger and invasive into adjacent tissue. [73, 82, 98] Malignant degeneration appears to be extremely rare, few cases have been reported. [100]

Benign Metastasizing Leiomyoma:
- Benign metastasizing leiomyoma is an ill-defined clinicopathological condition which features 'metastatic' histologically benign smooth muscle tumour deposits in the lung, lymph node or abdomen that appears to be derived from a benign uterine leiomyoma. [77, 82, 101]

- Most examples of "benign metastasizing leiomyomas" however appear to be either a primary benign smooth muscle lesion of the lung in a women with a history of uterine leiomyoma or pulmonary metastases from a histologically non-informative smooth muscle neoplasm of the uterus. [77] The findings of a recent cytogenetic study were most consistent with a monoclonal origin of uterine and pulmonary tumors and the interpretation that the pulmonary tumours were metastatic. [77] These tumors usually posses high content of oestrogen and progestogen receptors. [85]

SMOOTH MUSCLE TUMORS OF UNUSUAL MALIGNANT POTENTIAL (STUMP)

- A smooth muscle neoplasm that cannot be diagnosed reliably as benign or malignant on the basis of generally applied criteria. The terms "uncertain malignant potential" "low malignant potential" and "with limited experience" have been used. [73, 77, 82]

- The term low malignant potential is used for tumours that had coagulative tumor cell necrosis, MI<10 MFS/10HPF and absent to mild cytological atypia.

- Other example where the term has been used include tumors with only minimal atypia and a low mitotic index but uncertainty over type, and smooth muscle tumors with diffuse, severe atypia and a low mitotic rate but uncertainty about coagulative tumour cell necrosis.

- Most tumors in the uncertain malignant potential group probably have a favourable outcome but until further data becomes available the use of the term remains necessary. [73, 77, 82]

LEIOMYOSARCOMA

Epidemiology:
- Leiomyosarcoma are the most frequent pure sarcomas to arise in the uterus. [73] They account in most studies for between 25 to 45% of uterine sarcomas and upto 1% of all uterine malignancies. [73, 77, 82] The incidence of leiomyosarcoma is reported to be 0.3 – 0.4 /100.000 women per year. [77] Leiomyosarcoma arises exclusively in adults the median age of patients with leiomyosarcoma was 50-55 years in larger studies and 15% of patients were younger than 40 years. [77] Leiomyosarcomas tend to arise denovo and morphologically convincing examples of cases arising in leiomyomas are said to be rare. [73] The incidence of leiomyosarcoma in uterine leiomyomas is estimated to be between 0.13 to 0.29%. [102]

Clinical features:
- Clinical features are non specific; most common symptoms are abnormal vaginal bleeding, abdominal or pelvic pain, abdominal or pelvic mass and backache. [73, 77, 82, 103, 104, 105] Rapid growth of a uterine smooth muscle tumour in an elderly woman is very suspicious of leiomyosarcoma. [73]

Gross Appearance:
- Leiomyosarcomas are characteristically solitary intramural mass and are usually not associated with leiomyoma. Average diameter is 8 cm, fleshy in consistency and has variegated cut surface that is grey to white with areas of hemorrhage and yellow foci of necrosis. [73, 77, 82, 85, 104] In advanced cases evidence of extension beyond the uterus may be present. The margin may be macroscopically infiltrative and irregular. [73]

Histopathologic features:
- Leiomyosarcoma is a highly cellular tumour composed of fascicles of large spindled cells with marked pleomorphic nuclei and diffuse, moderate to severe atypia.

- Mitotic figures are typically numerous usually 20/10 HPF, and frequent atypical mitotic figures are often seen. Nuclear chromatin is typically coarse and nucleoli are prominent. Multi nucleated giant cells resembling osteoclasts are often seen in 65- 80% of leiomyosarcomas. Coagulative tumour cell necrosis is a common finding; irregular infiltrative border and vascular invasion are seen. [71, 73, 77, 82, 85]

- The diagnosis of leiomyosarcoma can be made in any tumour demonstrating coagulative tumor cell necrosis in the presence of cytologic atypia and any mitotic activity, as well as in tumours with an excess of 10 MF/10 HPF and coagulative tumor cell necrosis, in the absence of cytologic atypia. [76]

Epitheloid Leiomyosarcoma:
- Epitheloid leiomyosarcoma exhibits one of the patterns of epithelioid differentiation in addition to the usual features of leiomyosarcoma, i.e. cytologic atypia, tumour cell necrosis, high mitotic index. It has to be differentiated from benign epitheloid neoplasms. [77, 82, 106] It is suggested that lesions demonstrating increased mitotic activity, nuclear atypicality and infiltrative margins, especially those lacking clear cut encapsulation or circumferential hyalinization, be regarded as epithelioid leiomyosarcomas if the qualities of epithelioid characterization are fulfilled, even if the mitotic activity is less than five to ten HPFs. Biologic behaviour of this subset of unusual uterine mesenchymal tumours are, as yet, not well delineated, owing to the few reported cases in the literature. [106]

Myxoid Leiomyosarcoma:
- Myxoid leiomyosarcoma is a large, gelatinous neoplasm that usually appears circumscribed on gross examination. [77, 82, 107] Microscopically, the smooth muscle cells are usually widely separated by myxoid material. The characteristic low cellularity largely accounts for the presence of only a few mitotic figures per 10 HPF in most myxoid leiomyosarcomas. In almost all instances myxoid leiomyosarcomas show cellular pleomorphism and nuclear enlargement. They commonly show myometrial and sometimes vascular invasion. [77]

Clinicopathological correlation:

- Leiomyosarcoma spreads predominantly by the haematogenous route. The most common site of metastasis is lung. Other locations include bone, soft tissue, brain, kidney, spleen and ovary. [82, 105] Tumor size is a major prognostic factor, other prognostic factors suggested in various studies include, mitotic index, histologic grading, character of tumour border, vascular invasion, origin in a leiomyoma, extrauterine extension, patient age and uterine size. [108] Studies have also shown that grade of atypia, DNA ploidy, P53 index, AgNoR counts, percentage nuclear PCNA positivity, proliferative index on flow cytometry may have a role in predicting outcome in leiomyosarcoma. [73]

- Leiomyosarcoma is a highly malignant neoplasm; the variation in survival rates reported historically is largely the result of the use of different criteria for its diagnosis. Overall 5-years survival rate range from 15-20 %. 5 years survival rate is 40-70% in stage one and stage two tumours. Premenopausal women have a more favourable outcome in some studies; most recurrences are detected within two years. [77]

MISCELLANEOUS TUMORS OF MYOMETRIUM

Uterine Leiomyoma with Skeletal Muscle Differentiation
- Skeletal muscle differentiation within a leiomyoma is very rare; only two cases have been reported. [109] When it occurs in leiomyomas it can be considered as the phenotypic expression of the multiple differentiating potential of uterine smooth muscle cells which have a Mullerian origin. Nevertheless, skeletal muscle differentiation has also been observed in benign and malignant tumours which did not have muscular or Mullerian origin. Therefore, it may be a result of an aberrant genomic deregulation which occurs in neoplastic cells, regardless of their origin. [109]

Vascular Tumours and Malformations:
- True vascular tumours are rare, most frequently described appear to be capillary haemangiomas. Angiosarcomas are extremely rare, usually seen in pre and post menopausal women, have aggressive course. Presentation tends to be with bleeding and anemia. [73]

Adenomatoid Mesothelioma:
- It is a tumour derived from mesothelial cells usually an incidental finding in uteri removed for other causes. [77] They occur in perhaps 1% of uteri in reproductive age group and are benign [73, 77] Grossly, adenomatoid tumors are usually small tumours less than one cm in diameter, grey, rubbery, well circumscribed or ill defined in some cases. Most often found in the cornu and subserosally. [73, 77, 82]

Histological Appearance:
- These tumours are composed of small, irregular spaces lined by flat to cuboidal cells producing trabecular, microcystic, tubular arrays. The stroma contains large amount of smooth muscle and overall pattern closely mimics a leiomyoma with prominent lymphatics. The cytoplasm of mesothelial cells is abundant and nuclear atypia is absent. Immunohistochemically, the constituent mesothelial cells are strongly cytokeratin positive and no epithelial mucin is seen. [73, 77]

Stromal Nodule within a Uterine Leiomyoma:

- A yellow nodule within a uterine leiomyomas is usually due to the presence of mature adipose cells, diagnostic of a lipoleiomyoma. A case of leiomyoma associated with stromal nodule has been reported. [110] The yellow colour of the stromal nodule is due to fat accumulation, mostly within foam cells. Immunohistochemical analysis suggests a histocytic nature of foam cells. [110]

Mixed Endometrial stromal and smooth muscle tumors:

- There are tumours exhibiting prominent stromal and smooth muscle differentiation, recognizable by routine light microscopy, sometimes referred to as stromomyomas. [111]

- The clinical presentation of uterine mixed endometrial stromal smooth muscle tumors does not differ from that of uterine tumours of pure endometrial stromal or smooth muscle origin. Grossly, intramural, subserosal or submucosal in location, multinodular with either alternating soft, torn-yellow or white, or firm nodules or soft yellow modules embedded in or at the periphery of white, firm "leiomyomatus" tissue.

- Microscopically, endometrial stromal component is identical to that seen in typical endometrial stromal tumours, composed of closely packed small cells and numerous arterioles. Smooth muscle component occurs immediately adjacent to the stroma was characterized by nodules of variable size with prominent central hyalinization and thin bundles of collagen radiating towards the periphery, in which the tumour cells ere embedded, referred to as a "starburst" pattern. This pattern is not unique to mixed stroma and smooth muscle tumours. It also has been described in leiomyomas of the deep soft tissue, [112] in neurilemomas, [113] and hyalinizing spindle cell tumors with gaint sorettes. [114]

Pure Heterologous Sarcomas:
- These tumours are rare and there is debate as to whether many represent pure sarcomas or simply overgrowth of the sarcomatous component of carcinosarcomas or adenosarcomas. Most reported have been adult type rhabdomyosarcomas, chondrosarcomas and ostoesarcomas, though alveolar rhabdomyosarcomas, liposarcomas, alveolar soft part sarcomas, amongst others are also described. [73]

Uterine Myxoid Leiomyosarcoma within a Leiomyoma:
- Myxoid degeneration in benign leiomyomas is a well known phenomenon, but myxoid leiomyosarcoma arising in a leiomyomas is very rare and a case has been reported. [115]

- The case presented as gelatinous nodules within an otherwise leiomyoma. On microscopic examination, the gelantinous nodules were ill defined and showed cellular infiltrating leiomyosarcoma and were confirmed by MIB immunostain and P53 immuno stain. [115]

Malignant Mesenchymoma of the Uterus, arising in a leiomyomas:
- Malignant transformation of benign uterine leiomyomas may rarely occur. A case of malignant mesenchymoma of the uterus, arising in a leiomyoma has been reported. [116]

- Leiomyoma showed differentiation along multiple mesenchymal pathways, thus filling the criteria for a malignant mesenchymoma. Within the tumor osteosarcomatus, leiomyosarcomatus and adipose differentiation were seen, in addition to a morphologically benign leiomyomatous component.

- The relationship of the different components of the tumour was analysed by immunohistochemistry and with molecular analysis. [116]

Metastatic Carcinoma in Leiomyoma:

- Malignancies metastasizing to a uterine leiomyomas are very rare, few cases have been reported. The primary sites in these cases were breast, stomach, lung, pancreas and gall bladder.

- It is suggested that metastasis to leiomyomas may be considered in patients with primary carcinoma who show enlarging uterine tumours. [117, 118]

REFERENCES

1. Kurman RJ. Textbook of Blaustein's Pathology of Female Genital Tract. 5th edition, New York: Springer Verlag, 2002. Chapter no. 9, p. 235-227.
2. Bayer SR, DeCherney AH. Clinical Manifestaations and Treatment of Dysfunctional Uterine Bleeding. JAMA. 1993; 269: 1823-1828.
3. Sharma JB. Dysfunctional Uterine Bleeding (DUB). Obstetrics and Gynecology Today. 2000; 5 (11): 20-25.
4. Davey DA. Dewhurst's Textbook of Obstetric and Gynecology for Postgraduates. Glsgow: Blackwell Science. 1997, p. 590-608.
5. Hellweg GD. Histopathology of the Endometrium. 3rd edition, Berlin: Springer Verlag. 1981, p. 94-146.
6. Bereks JS, Rinehart RD, Adashi EY. Berek and Novaks Gynecology. 14th edition, Lippincot Willams and Wilkins. 2007, p. 149:160.
7. Purandhare CN. Dysfunctional Uterine Bleeding – An Update (FOGSI). New Delhi: Jaypee Brothers Medical Publishers. 2004.
8. Scommegna A., Dmowski PW. Dysfunctional uterine bleeding. Clin Obstet Gynecol. 1973; 16(3): 221-253.
9. Berek JS. Textbook of Novak's Gynecology. 12th edition, Maryland William and Wilkins. 1996. Chapter no. 7, p. 149-174.
10. Charles Mitchell. Regulation of the menstrual cycle. 6th edition, Maryland William and Wilkins. 1999. Chapter no. 6, p. 201-246.
11. Fox H. Haines and Taylor obstetrical and Gynecological Pathology. 4th edition, Churchill Livingstone. 1995. Chapter no. 10, p. 365-382.
12. Hellweg GD. Histopathology of the Endometrium. 3rd edition, Berlin: Springer Verlag. 1981, p. 50-80.
13. Symmers W. St. C. Female reproductive system: Systemic Pathology. Edinburgh. 1991, p. 129-145.
14. Kurman RJ. Textbook of Blaustein's Pathology of Female Genital Tract. 5th edition, New York: Springer Verlag, 2002. Chapter no. 9, p. 384-406.
15. Galle PC, McRae MA. Abnormal uterine bleeding.Finding and treating the cause. Postgrad Med. 1993; 93: 73-81.
16. Speroff L, Glass RH, Kase NG. Clinical Gynecologic endocrinology and infertility. 6th edition, Lippincott Williams & Wilkins. 1999.

17. Goldfarb JM, Little AB. Abnorml vaginal bleeding. N Engl J Med. 1980; 302: 474-488.
18. Povey WG. Abnormal uterine bleeding at puberty and climacteric. Clin Obstet Gynecol. 1970; 13: 474-488.
19. Stenchever MA, Droegemueller W, Herbst AL, Mishell DR. Comprehensive gynecology. 4th edition, St. Louis. Mosby. 2001, p. 155–177.
20. Rock JA, Jones HW. Te Linde's operative gynecology. 9th editon, Philadelphia: Lippincott Williams & Wilkins. 2003, p. 457-481.
21. Kilbourn CL, Richards CS. Abnormal uterine bleeding. Diagnostic considerations, management options. Postgrad Med. 2001; 109:137-147.
22. Wren BG. Dysfunctional uterine bleeding. Aust Fam Phys. 1998; 27: 371–377.
23. Baitlon O, Hadley JO. Endometrial biopsy. Pathologic findings in 3,600 biopsies from selected patients. Am J Clin Pathol. 1975; 63: 9–15.
24. Nickelsen C. Diagnostic and curative value of uterine curettage. Acta Obstet Gynecol Scand. 1980; 65: 693–697.
25. Van Bogaert L-J, Maldague P, Staquet J-P. Endometrial biopsy interpretation. Shortcomings and problems in current gynecologic practice. Obstet Gynecol. 1978; 51: 25–28.
26. Rubin SC. Postmenopausal bleeding: Etiology, evaluation, and management. Med Clin N Am. 1987; 71: 59–69.
27. Moghal N. Diagnostic value of endometrial curettage in abnormal uterine bleeding - A histopathological study. J Pak Med Assoc. 1997; 47: 295–299.
28. Scurry J, Brand A, Sheehan P, Planner R, Highgrade endometrial carcinoma in secretory endometrium in young women. A report of five cases. Gynecol Oncol. 1996; 60: 224–227.
29. Michael T. Mazur, Robert J. Kurman. Diagnosis of endometrial biopsies and curettings: A practical approach. 2nd edition, New York. Springer Science Buseness Media, Inc. 2005, p. 108-119.
30. Choo YC, Mak KC, Hsu C, Wong TS, Ma HK. Postmenopausal uterine bleeding of nonorganic cause. Obstet Gynecol. 1985; 66: 225–228.
31. Kiviat NB, Wolner-Hanssen P, Eschenbach DA, Wasserheirt JN, Paavonen JA, Bell TA et al. Endometrial histopathology in patients with culture-

proved upper genital tract infection and laparoscopically diagnosed acute salpingitis. Am J Surg Pathol. 1990; 14: 167–175.

32. Paavonen J, Aine R, Teisala K, Heinonen PK, Punnonen R. Comparison of endometrial biopsy and peritoneal fluid cytologic testing with laparoscopy in the diagnosis of acute pelvic inflammatory disease. Am J Obstet Gynecol. 1985; 151: 645–650.

33. Greenwood SM, Moran JJ. Chronic endometritis. Morphologic and clinical observations. Obstet Gynecol 1981; 58: 176–184.

34. Juan Rosai MD. The text book of Rosai and Ackerman's Surgical Pathology, 10th edition, 2011 (Vol. 2), Ch-19, p. 1477-1518.

35. Rotterdam H. Chronic endometritis. A clinicopathologic study. Pathol Annu. 1978; 13: 209–231.

36. Buckley CH, Fox H. Biopsy pathology of the endometrium. 2nd edition, London: Arnold. 2002.

37. Kurman RJ. Blaustein's pathology of the female genital tract. 5th edition, New York: Springer-Verlag. 2002, p. 421–466.

38. Winkler B, Reumann W, Mitao M, Gallo L, Richart RM, Crum CP. Chlamydial endometritis. A histological and immunohistochemical analysis. Am J Surg Pathol. 1984; 8: 771–778.

39. Bazaz–Malik G, Maheshwari B, Lal N. Tuberculous endometritis. A clinicopathological study of 1000 cases. Br J Obstet Gynaecol. 1983; 90: 84–86.

40. Nogales-Ortiz F, Tarancon I, Nogales FF. The pathology of female genital tuberculosis. A 31- year study of 1436 cases. Obstet Gynecol. 1979; 53: 422–428.

41. Schaefer G, Marcus RS, Kramer EE. Postmenopausal endometrial tuberculosis. Am J Obstet Gynecol. 1972; 112: 681–687.

42. Robboy SJ, Mutter GL, Prat J, Bentley RC. Robboy's Pathology of the Female Genital Tract. 2nd edition, Churchill Livingstone Eslevier. 2009, p. 361-364.

43. Savelli L et al. Histopathologic features and risk factors for benignity, hyperplasia and cancer in endometrial polyps. Am J Obstet Gynecol. 2003; 188: 927-30.

44. Kin KR, Peng R, Robboy SJ. A diagnostically useful histopathologic feature of endometrial polyp. Am J Surg Pathol. 2004; 28(8): 1057- 1061.

45. Kelley P, Dobbs SP, McCluggage WG. Endometrial hyperplasia involving endometrial polyps: Report of a series and discussion of the significance in an endometrial biopsy specimen. BJOG. 2007; 114: 944- 950.
46. Van Bogaert L-J. Clinicopathologic findings in endometrial polyps. Obstet Gynecol. 1988; 71: 771–773.
47. Hellweg GD. The endometrium of infertility. A review. Pathol Res Pract. 1984; 178: 527–537.
48. Bakour SH, Gupta JK, Khan KS. Risk factors associated with endometrial polyps in abnormal uterine bleeding. Int J Gynaecol Obstet. 2002; 76: 165–168.
49. Cohen I, Azaria R, Bernheim J, Shapira J, Beyth Y. Risk factors of endometrial polyps resected from postmenopausal patients with breast carcinoma treated with tamoxifen. Cancer. 2001; 92: 1151–1155.
50. Fukunaga M, Endo Y, Ushigome S, Ishikawa E. Atypical polypoid adenomyomas of the uterus. Histopathology. 1995; 27: 35–42.
51. Kurman RJ. Blaustein's pathology of the female genital tract. 5th edition, New York: Springer-Verlag. 2002, p. 440.
52. Whitehead MI, Fraser D. The effects of estrogens and progestogens on the endometrium. Obstet Gynecol Clin North Am. 1987; 14: 299–320.
53. Goldstein SR. The effect of SERMs on the endometrium. Ann NY Acad Sci 2001; 949: 237–242.
54. Spitz IM, Bardin CW. Mifepristone (RU 486)—a modulator of progestin and glucocorticoid action. N Engl J Med. 1993; 329: 404–412.
55. Kurman RJ. Blaustein's pathology of the female genital tract. 5th edition, New York: Springer-Verlag. 2002, p. 467-500.
56. Soslow RA, Pirog E, Isacson C. Endometrial Intraepitelial Carcinoma with associated peritoneal cercinomatosis. Am J Surg Pathol. 2000; 24: 726-732.
57. Robboys SJ, Mutter GL. Robboy's pathology of female reproductive tract. 2nd edition, Churchill Livingstone Elsevier. 2009. Chapter No. 16, p. 396-415.
58. Fletcher DM. Diagnostic histopathology of tumours. 3rd edition, Vol.1, Elsevier Churchill Livingstone. 2007, p. 653.
59. Kurman RJ. Blaustein's pathology of the female genital tract. 5th edition, New York: Springer-Verlag. 2002, p. 501-535.

60. Manisha Ram, Minakshi Bharadwaj, Rajbala Yadav. Endometrial intraepithelial carcinoma: A case report and brief review. Indian J Path & Microbiol. 2008; 51(4): 512-514.
61. Robboys SJ, Mutter GL. Robboy's pathology of female reproductive tract. 2nd edition, Churchill Livingstone Elsevier. 2009, p. 396- 415.
62. McCluggage WC. Malignant biphasic tumours of uterus: Carcinosarcomas or metaplastic carcinomas? J Clin Pathol. 2002; 55: 321-325.
63. Sarah E. Ferguson, Carmen Tornos, Amanda Hummer, Richard R. Barakat and Robert A. Soslow. Prognostic features of surgical stage I uterine carcinosarcoma. Am J Surg Pathol. 2007; 31(11): 1653- 1661.
64. Basel Altrabulsi, Anais Malpica, Michael T. Deavers, Diane C. Bodurka, Russell Broaddus et al. Undifferentiated carcinoma of the endometrium. Am J Surg Pathol. 2005; 29: 1316-1321.
65. Abdulmohsen Alkush, Zainab H. Abdul-Rahman, Peter Lim, Michael Scuzlzer, Andrew Coldman et al. Description of a novel system for grading of endometrial carcinoma and comparison with existing grading systems. Am J Surg Pathol. 2005; 29: 295-304.
66. Sharon Nofech-Mozes, Zeina Ghorab, Nadia Ismiil, Ida Ackerman, Gillian Thomas et al. Endometrial endometrioid adenocarcinoma. Am J Clin Pathol. 2008; p. 129, 110-114.
67. Soslow RA, Bissonnette JP, Andrew Wilston, Ferguson SE, Kaled M. Alektiar, et al. Clinicopathologic analysis of 187 high-grade endometrial carcinomas of different histologic subtypes: Similar outcomes Belie Distinctive Biologic Differences. Am J Surg Pathol. 2007; 31: 979-987.
68. Mazur MT, Kurman RJ. Diagnosis of endometrial biopsies and curretings: A practical approach. 2nd edition, New York. Springer Science Buseness Media Inc. 2005, p. 34-99.
69. Jordan LB, Al-Nafussi A, Beattie G. Cotyledonoid hydropic intravenous leiomyomatosis: A new variant leiomyoma. Histopathology 2002; 40: 245- 52.
70. Roth LM, Reed RJ, Sternberg WH. Cotyledonoid dissecting leiomyoma of the uterus the Sternberg tumour. Am J Surg Pathol 1996; 20(12): 1455-61.
71. Wilkinson N, Rollason TP. Recent advances in the pathology of smooth muscle tumours of the uterus. Histopathology 2001; 39: 331-41.

72. Vollenhoven BJ, Lawrence AS, Healy DL. Uterine fibroids: A clinical review. Br J Obstet Gynaecol 1990 April; 97: 285-98.
73. Rollason TP, Wilkinson N. Non neoplastic condition of the myometrium and pure mesenchymal tumours of the uterus. In: Obstetrical and GynaecologicalPathology, Fox H, Wells M, eds, 5th edition, vol. 2, New York: Churchill Livingstone, 2003.
74. Uterus. In: The Gray's Anatomy, Anatomical Basis of Clinical Practice., Susan Standring ed., 39th edition, Elsevier, Churchil Livingstone, 2005, p. 1331.
75. Frederick KT. The myometrium. In Gynecologic Pathology. 2001, p. 223.
76. Bell SW, Kempson RL, Hendrickson MR. Problematic uterine smooth muscle neoplasms: A clinicopathologic study of 213 cases. Am J Surg Pathol 1994; 18: 535-58.
77. Hendrickson MR, Tavassoli FA, Kempson RL, McCluggage WG, Haller U, Kubik-Huch RA. Mesenchymal tumors and related lesions. In: TavassoliFA, Deville P, eds. World Health Organization of Tumors. Pathology and Genetics of Tumors of the Breast and Female Gynetical Organs. France: IARC Press, 2003, p. 80.
78. Hains M, Taylor CD. Fibromyoma. In: Gynaecological Pathology, 2nd edition, New York: Churchill Livingstone, 1975.
79. Kawaguchi K, Fujii S, Konishi T, Nanbu Y, Nonogaki H, Mori T. Mitotic activity in uterine leiomyomas during the menstrual cycle. Am J ObstetGynecol 1989 March; 160(3): 637-41.
80. Cramer SF, Patel A. The non random regional distribution of uterine leiomyomas: A clue to histogenesis? Hum Pathol 1992 June; 23(6): 635-38.
81. Chhabra S, Jaiswal M. Vaginal management of uterocervial myomas. J of Obstet and Gynaecol of India 1996; 46: 260-63.
82. Zalovdek C, Hendrickson M. mesenchymal tumors of the uterus. In Kurman RJ ed. Blausteins Pathology of the female genital tract. 5th edition, Springer, 2002.
83. Chhabra S, Ohri N. Leiomyomas of uterus – A clinical study. J of Obstet and Gynaecol of India 1993; 43(3): 437-39.
84. Clement PB, Young RH, Scully RE. Diffuse, pernodular, and other patterns of hydropic degeneration within and adjacent to uterine leiomyomas. Am J Surg Pathol 1992; 16(1): 26-32.

85. Leiomyomas. The Uterine Corpus. In: Hendrickson MR, Congracre TA, Kempson RL eds. Sternbergs diagnostic surgical pathology, 4th edition, Lippincott, Williams and Wilkins, 2004.
86. O'Connor DM, Norris HJ. Mitotically active leiomyomas of the uterus. Hum Pathol 1990 Feb; 21: 223-27.
87. Perrone T, Dehner LP. Prognostically favorable "mitotically active" smooth muscle tumours of the uterus: A clinicopathological study of ten cases. Am J Surg Pathol 1988; 12(1): 1-8.
88. Oliva E, Young RH, Clement PB, Bhan AK, Scully RE. Cellular benign mesenchymal tumours of the uterus: A comparative morphologic and immunohistochemical analysis of 33 highly cellular leiomyomas and six endometrial stromal nodules, tow frequently confused tumours. Am J Surg Pathol 1995; 19(7): 757-68.
89. Myles JL, Hart WR. Apoplectic leiomyomas of the uterus: A clinicopathologic study of five distinctive haemorrhagic leiomyomas associated with oral contraceptive usage. Am J Surg Pathol 1985; 9: 789-805.
90. Downes KA, Hart WR. Bizarre leiomyomas of the uterus: A comprehensive pathologic study of 24 cases with long term followup. Am J Surg Pathol 1997; 22(11): 1261-70.
91. Kurman RJ, Norris HJ. Mesenchymal tumours of the uterus: Epitheloid smooth muscle tumors including leiomyoblastoma and clear cell leiomyoma - A clinical and pathological analysis of 26 cases. Cancer 1976; 37: 1853-65.
92. Prayson RA, Goldblum JR, Hart WR. Epitheloid smooth muscle tumours of the uterus: A clinicopathologic study of 18 patients. Am J Surg Pathol 1997; 21(4): 383-391.
93. Hyde E, Geisinger KR, Marshall RB, Jones TL. The clear cell variant of uterine epithelial leiomyoma: An immunologic and ultrastructural study. Arch Pathol Lab Med 1989; 113: 551-53.
94. Setna Z, Siddiqui MS, Hussainy AS, Muzaffar S, Hasan SH. Pure lipoma of the uterus: An extremely rare entity. Indian J Pathol Microbiol 1999 July; 42(3): 383-84.
95. Fukunaga M, Usuigome S. Dissecting leiomyoma of the uterus with extrauterine extension. Histopathology 1998; 32: 160-64.

96. Roth CM, Reed RJ. Dissecting leiomyoma of the uterus other than coteledonoid dissecting leiomyoma: A report of eight cases. Am J Surg Pathol 1999; 23: 1032-1039.
97. Norris HJ, Maj TP. Mesenchymal tumours of the uterus, intravenous leiomyomatosis: A clinical and pathological study of 14 cases. Cancer 1975; 36: 2164-78.
98. Mulvany NJ, Ostor AG, Ross I. Diffuse leiomyomatosis of the uterus. Histopathology 1995; 27: 175-79.
99. Clement PB. Intravenous leiomyomatosis of the uterus. In Pathology Annual 1988; 23: 153-83.
100. Akkersdijk GJM, Flu PK, Giard RWM, Lent MV, Wallenburg HCS. Malignant leiomyomatosis peritonealis disseminate. Am J Obstet Gynecol 1990; 163: 591-3.
101. Cramer SF, Meyer JS, Kraner JF, Camel M, Mazur MT, Marilyn S et al. Metastasizing leiomyoma of the ulcers. S-phase fraction, estrogen receptor and ultrastructure. Cancer 1980; 45: 932-37.
102. Leibsohn S, Gerrit Dablaing, Mishell DR, Schlaerth JB. Leiomyomas in a series of hysterectomies performed for presumed uterine leiomyomas. Am J Obstet Gynecol 1990; 162: 968-76.
103. Parker WH, Yao Shifu, Berek JS. Uterine sarcoma in patients operated on for presumed leiomyoma and rapidly growing leiomyoma. Obstet Gynecol 1994; 83: 414-8.
104. Aaro CA, Symmonds RE, Dockerty MB. Sarcoma of he uterus 1966 Jan; 94: 101-09.
105. Hart WR, Billman JK. A reassessment of uterine neoplasms originally diagnosed as leiomyosarcomas. Cancer 1978; 41: 1902-10.
106. Buscema J, Carpenter SE, Rosenshein B, Woodruff JD. Epithelioid leiomyosarcomas of the uterus. Cancer 1986; 57: 1192-96.
107. Pounder DJ, Iyer PV. Uterine leiomyosarcomas with myxoid stroma. Arch Pathol Lab Med 1985; 109: 762-64.
108. Evans HL, Chawla SP, Simpson C, Finn KP. Smooth muscle neoplasms of the uterus other than ordinary leiomyoma: A study of 46 cases, with emphasis on diagnostic criteria and prognostic factors. Cancer 1988; 62: 2239-47.

109. Fornelli A, Pasquinelli G, Eusebi V. Leiomyoma of the uterus showing skeletal muscle differentiation: A case report. Hum Pathol 1999 March; 30: 356-59.
110. Marbaix E, Pirard C, Noboa P. A bright yellow stromal nodule within a uterine leiomyoma: A case report. Histopathology 39; 646-48.
111. Oliva E, Clement PB, Young RH, Scully RE. Mixed endometrial stromal and smooth muscle tumours of the uterus: A clinicopathologic study of 15 cases. Am J Surg Pathol 1998; 22(8): 997-1005.
112. Kilpatrick SE, Mentzel T, Fletcher CDM. Leiomyomas of soft tissue: Clinicopathologic analysis of a series. Am J Surg Pathol 1994; 18: 576-82.
113. Enzinger FM, Weiss SW. Benign tumors of peripheral nerves. In Enzinger FM, Weiss SW, eds. Soft tissue tumors. 3rd edition, Louis, MO: Cumosby 1995; p. 821-88.
114. Lane KL, Shannon RJ, Weiss SW. Hyaliizing spindle cell tumour with giant rosettes: A distinctive tumor closely resembling low grade fibromyxoid sarcoma. Am J Surg Pathol 1997; 21: 1481-8.
115. Mittal K, Popiolek D, Demopoulos RI. Uterine myxoid leiomyosarcomas within a leiomyoma. Hum Pathol 2000 March; 31: 398-400.
116. Bakker MA, Hegt VN, Sleddens HBFM, Nuijten ASM, Dinjens WNM. Malignant mesenchymoma of the uterus, arising in a leiomyoma. Histopathology 2002; 40: 65-70.
117. Lanjewar DN, Shetty CR. Metastatic carcinoma in uterine leiomyoma. Indian J Pathol Microbiol 1997; 40(3): 409-411.
118. Burton JL, Wells M. Tumors of the myometrium. In Fletcher CDM ed. Diagnostic histopathology of tumors. 2nd edition, Vol. 1, Churchill Livingstone, Hartcourt Publishers, 2000.

www.ingramcontent.com/pod-product-compliance
Lightning Source LLC
Chambersburg PA
CBHW080918170526
45158CB00008B/2153